CAMBRIDGE STUDIES IN MODERN BIOLOGY: 1

COASTAL LAGOONS

The natural history of a neglected habitat

R. S. K. BARNES
St Catharine's College & Department of Zoology, University of Cambridge

COASTAL LAGOONS

The natural history of a neglected habitat

CAMBRIDGE UNIVERSITY PRESS
Cambridge
London New York New Rochelle
Melbourne Sydney

CAMBRIDGE UNIVERSITY PRESS
Cambridge, New York, Melbourne, Madrid, Cape Town, Singapore, São Paulo, Delhi

Cambridge University Press
The Edinburgh Building, Cambridge CB2 8RU, UK

Published in the United States of America by Cambridge University Press, New York

www.cambridge.org
Information on this title: www.cambridge.org/9780521299459

First published 1980
Re-issued in this digitally printed version 2009

A catalogue record for this publication is available from the British Library

ISBN 978-0-521-23422-1 hardback
ISBN 978-0-521-29945-9 paperback

In memory of Dr Charlie Boyden, friend, colleague and companion on many field trips to lagoons

and in honour of Slartibartfast, maker of coastlines

CONTENTS

PREFACE

To my knowledge, this is the first book in English to review the natural history of the world's coastal lagoons. Therefore I have not had the benefit of reviewers' impressions of other people's failings when attempting this subject: I offer this small work in the hope that others will now have, and profit by, that privilege.

Lagoons are an important part of our planet's coastal scene, yet lagoonal biology – like lagoonal science in general – is in its infancy. I do not pretend to know the reason for this neglect, but I would like to hope that the following pages will go some way towards introducing and commending these fascinating habitats to a wider audience. Although a biologist should perhaps be content to confine himself to biology, I have presumed to attempt a more complete canvas than that to which my competence might entitle me. In mitigation I plead the impossibility of an understanding of the nature of the real world through parochial disciplinary approaches: one can only appreciate and investigate details of the part when one has a clear image of the whole. Whether or not I have succeeded in creating a clear image of the lagoonal whole, I cannot be the judge; such at least has been my intent. I also take the point of the teaching of Callimachus of Cyrene that a large book is a large evil; accordingly I have endeavoured to be as concise as the constraint of readability will allow.

In some measure, this volume can be regarded as a companion to my earlier one on estuaries (*Estuarine Biology*; Arnold, 1974); in any event, I have tried to avoid duplication of information. As will be evident from the following pages, lagoons vary from minute pools within shingle formations which fill with sea water at high tide and drain on the ebb to large areas of shallow coastal sea sheltered by a chain of barrier islands, and their contained water ranges from almost fresh to ionic concentrations three times that of sea water. Nevertheless, as appropriate to a general introductory account, I have endeavoured to generalise wherever possible, to concentrate on common properties and features, and to restrict the bibliography to reviews, 'key papers' and some of the more recent articles to have appeared.

Lastly, although written for the undergraduate and as an introduction for the postgraduate student, I very much hope that the book will be of some interest to my colleagues, whether amateur or professional.

Cambridge, St Swithin's Day, 1979 R.S.K.B.

ACKNOWLEDGEMENTS

As on previous occasions, I am deeply grateful to Dr Colin Little for his careful and constructive reading of the typescript. I am also most grateful to Dr Charles Goodhart for very kindly translating material from Russian into English for me.

I am indebted to the following authors and publishers for permitting me to reproduce illustrations of which they hold the copyright: The Royal Society of Victoria and Professor E. C. F. Bird (Fig. 2.8); Academic Press Inc. (London) Ltd and Dr R. R. C. Edwards (Fig. 3.2, 3.8*a*); Blackwell Scientific Publications Ltd and Dr C. Howard-Williams (Fig. 4.2); Springer-Verlag (Heidelberg) and Professor T. Fenchel (Fig. 5.4); Blackwell Scientific Publications Ltd (Fig. 7.3); and Mezhdunarodnaya Kniga, Moscow (Fig. 2.5).

1

Introduction

The *Oxford English Dictionary* allows two meanings of the word 'lagoon': an area of salt or brackish water separated from the adjacent sea by a low-lying sand or shingle barrier; and a lake-like stretch of water enclosed in a coral atoll. The term coastal lagoon refers to the first of these two situations.

That a volume written and published in England should be devoted to coastal lagoons may at first occasion some surprise as there are relatively few examples of this type of habitat in the British Isles. It has been estimated by Cromwell (1971), however, that 'features less than 10 m above sea level. . . which are impounding, or have at one time impounded, bodies of water between themselves and the mainland' occupy 13% of the world's coastline. Of this total, 34% is contributed by North America; Europe only has 5.3% of its coastline assignable to this category and is the continent with the lowest proportion of lagoonal coast (Table 1.1.).

Table 1.1. *Extent of the world's barrier/lagoonal coastline. From Cromwell, 1971*

Continent	Length (km) of barrier/ lagoonal coastline	% of continent's coastline so formed	% of world's lagoonal coast
North America	10765	17.6	33.6
Asia	7126	13.8	22.2
Africa	5984	17.9	18.7
South America	3302	12.2	10.3
Europe	2693	5.3	8.4
Australia	2168	11.4	6.8
Total	32038		

There are particularly extensive series of lagoons along the eastern and Gulf-of-Mexico coasts of the USA, in Mexico itself, in Brazil, West Africa, Natal, the southern and eastern shores of the Indian peninsula, south-west and south-east Australia, Alaska, Siberia, in the Landes region of France, around the Mediterranean (e.g. near Venice, in the Gulf of Lions and around the Nile Delta), and around the southern Baltic, Black and Caspian Seas (Fig. 1.1). All regions, it may be noted, with a low tidal range. In Britain, small lagoons are relatively common in East Anglia and in the south and south-west.

Comparatively easy as it may be to identify exceptional areas of lagoonal coastline, it is undoubtedly more difficult to decide which natural features may properly be considered lagoons. This results from the leeway permitted by the words 'separated' and 'impounded' in the definitions above. Because of the requirement for their water to be marine, or at least brackish, lagoons are rarely completely isolated from the sea. Characteristically, they have a channel (or series of channels) through which water is exchanged with the larger adjacent water body, although in some small lagoons exchange may be effected only by percolation through the confining barrier. On complete isolation, the contained water usually becomes fresh (e.g.

Fig. 1.1. World distribution of barrier/lagoonal coastlines. After Leont'ev & Leont'ev, 1957 and Gierloff-Emden, 1961.

Slapton Ley in Devon and several of the Landes lagoons) and the lagoon evolves into a freshwater coastal lake or pond (see p. 14).

Therefore lagoons are only semi-isolated systems, permitting differences of opinion on the extent of separation required for the granting of lagoonal status. Some authors, for example, describe the wadden areas delimited by the Frisian Island chain which extends along the south-eastern shore of the North Sea (Fig. 1.2*a*) as being lagoons, although effectively the waddens are only very sheltered, shallow regions of the North Sea and they are more reasonably regarded as part of the marine environment. The many, relatively large gaps between the barrier islands suffice to prevent the establishment of a separate system. The typical lagoon has an entrance channel which is very small in relation both to the size of the lagoon itself and to the length of the barrier (Fig. 1.2*b*), down to the limiting state of no discrete channel at all (Fig. 1.2*c*).

Essentially similar to the latter, although with no permanent connection with the sea, are the small lagoons entrapped by movements of sedimentary material in growing spits, deltas (Fig. 1.2*d*) or similar geomorphological features. These are relict systems which, whilst they persist, show resemblances both to lagoons and to pans on salt-marshes and lakes in dune systems which may only communicate with the sea on an irregular temporal basis. Man has created equivalent lagoons during the reclamation of coastal marshes (Hunt, 1971; Barnes & Jones, 1975).

Lagoons, therefore, as one might expect, grade into other coastal habitat types: into semi-enclosed marine bays, into freshwater lakes, and into estuaries; and some of these intergradations may represent stages in an evolutionary sequence (pp. 7–19). Any attempt at rigid compartmentalisation will therefore be to a large degree artificial and arbitrary. Although the typical estuary and the typical lagoon are very different, the boundary between various habitat types is here at its most insubstantial and the two environments share many ecological features. Several differences are characterised on pp. 20–53, but from a gross geographical viewpoint, the matter, once again, ultimately comes down to the relative width of the entrance/exit channel (and the volume of freshwater or tidal input, which of course are interrelated). Estuaries, although often partially enclosed by a shingle spit, have relatively wide mouths and a large exchange or through-put of

Fig. 1.2. *a.* The Dutch/German Waddenzee (as it was before enclosure of the Zuiderzee) partially isolated from the North Sea by the Frisian Island chain. *b.* A lagoon with a single, narrow entrance channel: a composite diagram of the lagoons occurring on the shores of the Chukchi Sea, Eastern Siberia. Length may be 2–30 km. *c.* Isolated lagoons near Novorossiysk on the Black Sea; in general shape they are typical of the lagoons around the Eurasian Inland Seas. *d.* Isolated ephemeral lagoons in the Ebro Delta, eastern Spain, after Ferrer & Comin, 1979.

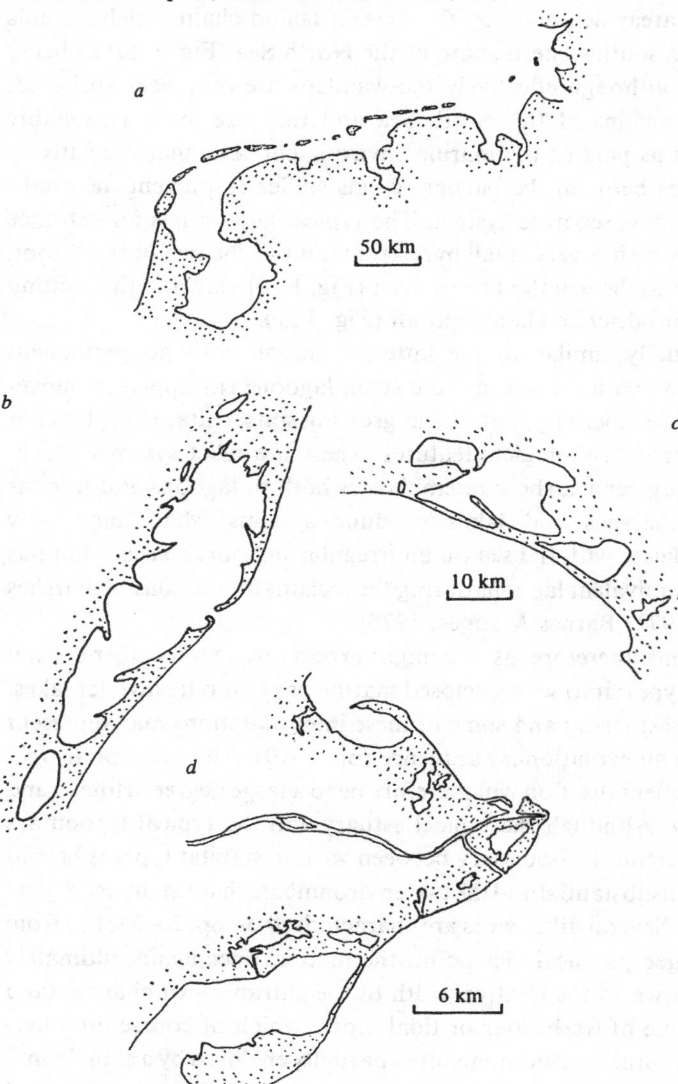

Fig. 1.3 *a–c*. Lagoon systems enclosed by offshore barrier-island chains in the USA: *a*, Long Island, New York; *b*, Pamlico and Albemarle Sounds, N. Carolina; *c*, the Laguna Madre and associated lagoons, Texas. *d*. A lagoon enclosed within the land: the Lagoa dos Patos between Porto Alegre and Rio Grande, Brazil.

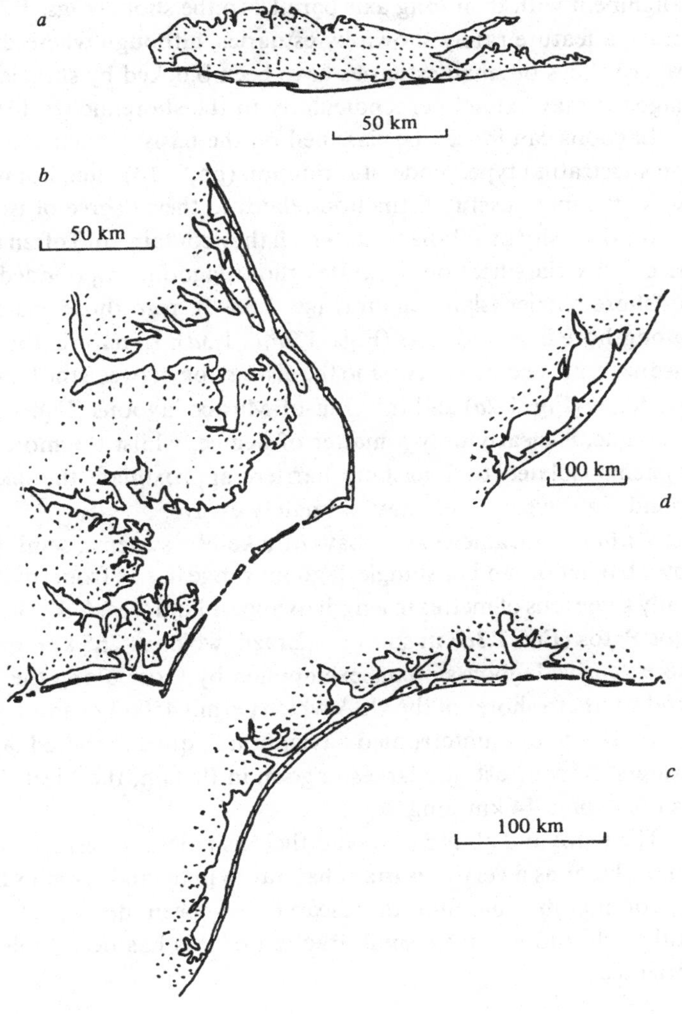

water in relation to their volume, whilst lagoons have relatively small mouths and little exchange or through-put. A number of geographers have distinguished between 'estuarine lagoons' into which rivers discharge and 'marine lagoons' without a major, running freshwater input, but biologically this distinction does not really seem to be warranted. A further characteristic of lagoons is their frequent alignment with their long axis parallel to the shore (Figs. 1.2*b*, 1.3*a*, *c*, *d*), a feature rarely shown by estuaries, although where drowned river valleys or small estuaries have been blocked by shingle ridges, lagoons may extend perpendicularly to the shoreline (p. 13).

Lagoons can indeed be classified on the basis of their salinity, or on substratum type, mode of formation (pp. 7–14), etc., but in many ways the most useful distinction relates to their degree of isolation, to the diversity of sub-habitats which they contain, and often to their size. This classification separates those lagoons impounded by an offshore barrier-island chain (Figs. 1.3*a*–*c*) from those enclosed by alongshore bars and spits (Figs. 1.2*b*,*c*, 1.3*d*). In nature, the former are nearer to enclosed bays, and the distinction between the North-Sea waddens (Fig. 1.2*a*) and the Gulf-of-Mexico lagoons (Fig. 1.3*c*), for example, is clearly only a matter of degree, whilst the more inland systems isolated by longshore barriers approximate the lakes and ponds into which they may ultimately evolve.

Within this framework of bay- or lake-like systems, semi-isolated by a barrier of sand or shingle, lagoons range in size from small pools only some tens of metres in length to huge inland 'seas' like the Lagoa dos Patos (Fig. 1.3*d*) in southern Brazil, which is 265 km long, and to extensive lagoon-systems exemplified by those along the eastern and southern shores of the USA which permit 4500 km and 1000 km, respectively, of uninterrupted navigation in quiet sheltered lagoonal waters. In contrast, the largest lagoon in Britain, the Fleet, Dorset, is only some 14 km long.

Their size and abundance, and their fisheries interest (pp. 69–70), make lagoons a very important habitat to man, and, even in Britain, a considerable amount of research has been devoted to them, although, sadly, only a small fraction of this has been biologically oriented.

2

Formation and fate of lagoons

Lagoons, especially the smaller ones, are rapidly changing, highly dynamic systems and their biology cannot be understood except within the framework of their formation, evolution and subsequent decline. Accordingly, this chapter will investigate their geographical and geological history. It will be evident from the previous chapter that the story of a lagoon is intimately connected with the barrier enclosing it – one cannot have one without the other – and hence we must start by looking at the barriers themselves.

2.1 Enclosure of lagoons by offshore barriers

Although many barriers have had a composite origin, it is convenient to consider the formation of offshore and longshore barriers separately. Offshore bars and barriers are produced by wave action on shallow, gently shelving sand coasts (or those comprising mixed sands and pebbles). Wave action may be of two types, constructive and destructive. Constructive waves plunge obliquely up a beach, generating a powerful swash but a relatively small backwash; destructive waves, on the other hand, plunge more vertically downward, creating a small swash but a powerful backwash. Constructive waves therefore tend to move sediment up an incline, whilst destructive waves move material down slope (see, e.g., King, 1972, for further details). When constructive waves enter shallow water they begin to break and their shoreward motion and swash move particles of sediment towards the coast, and when waves begin to break some distance offshore, a bar is formed there by this deposited sediment. Sand is, as it were, swept from offshore deposits in lines parallel to the coast which accumulate and merge in the breaking zone.

During the last 15000 years or so, sea level has risen some 100 m (although changes in the last 5000 yr have been minor). This has had the effect not only of drowning low-lying areas, but also of permitting the sweeping action of constructive waves gradually to roll large quantities of material up the gentle slopes so that the barriers contain more material than they would have done had sea level been constant during this period. It is also possible that some barriers have received sand blown landward by wind during times of falling sea level (i.e. some 60000 yr ago). Once the collected material is sufficient to extend above the water level, it can also, e.g., during low tide in relatively tidal areas, receive additional quantities of sand wind-blown from the intertidal zone so that sand-dunes may develop on the crest.

The length of the barriers will depend on shoreline and subtidal topography, but very long barriers may be formed – Padre Island, enclosing the Laguna Madre in Texas, is 200 km long, and a 600 km barrier, admittedly with some gaps in it, extends along the western coast of Kamchatka. Sand is the most common barrier material, though shingle, which because of its weight is less easily moved, does participate in some offshore barriers (especially those formed by storm waves). It is more abundant, however, in the longshore barriers considered below.

If offshore barriers are long enough to seal or effectively seal shallow coastal bays, lagoons are thereby formed from the erstwhile intertidal and shallow subtidal regions. A good example of the incipient enclosure of a lagoon by such a barrier may be seen off the North Norfolk coast between Holkham and Wells.

2.2 Enclosure of lagoons by longshore barriers

If the coast is steeply sloping, waves will break on the beach and not offshore. Waves also do not always approach parallel to a beach, and the effect of striking a beach at an angle is to move material along it. The swash of constructive waves approaching at an angle will move water and sand up the shore at the angle of attack, but the backwash will be directed, by gravity, perpendicularly to the shoreline, creating a longshore drift of sand (or shingle) (Fig. 2.1).

Particularly importantly for the cases under consideration here, material moved along a coast tends to maintain the alignment of that stretch of coast, even though the shoreline itself may change orientation. Thus spits form out across inlets or bays of the sea (Fig. 2.2*a*) and these spits, like offshore barriers, may be very long (spits in the Black and Caspian Seas achieve lengths of 70–100 km) and may seal off bays to form lagoons (Fig. 2.2*b*). Of course, material may move along offshore barriers, thereby joining together barrier islands formed separately, in exactly the same way and offshore barriers may

Fig. 2.1. Diagram to illustrate the process of longshore drift.

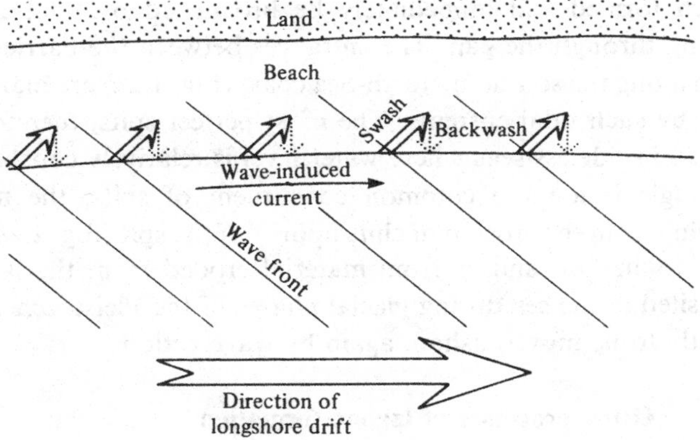

Fig. 2.2. *a*. Semi-diagrammatic representation of the northern spit partially enclosing Poole Harbour, Dorset, showing the alignment of the spit in relation to that of the coastline along which material is drifting. *b*. Spits enclosing lagoons on the Baltic-Sea coasts of Poland, the Russian SFSR and the Lithuanian SSR.

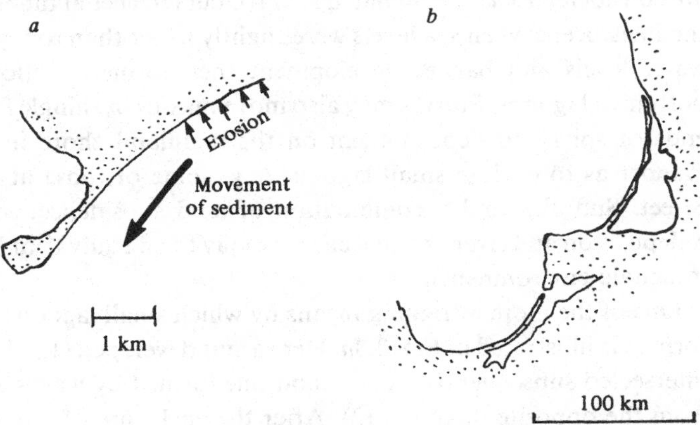

become attached to the coast at headlands, etc. So compositely formed barriers are common.

It is here that the importance of tidal range, mentioned on p. 2, can be appreciated. Barriers, being constructed by wind-driven waves and currents, do not require tidal forces for their generation: indeed the opposite is often the case. Large tidal forces resulting from large tidal ranges can create breaches in sedimentary barriers and any breach formed will immediately be broadened by tidal currents rushing through the gap. The entrances between the barrier-island chain along the southern North-Sea coast (Fig. 1.2*a*) are maintained open by such tidal currents. The most perfect spits, therefore, are formed in tideless seas where water level is relatively constant.

Shingle is a more common constituent of spits, the material deriving from the erosion of cliffs updrift of the spit (Fig. 2.2*a*), from river discharges, and/or from material eroded from the land and deposited in the sea during glacial phases of the Pleistocene, subsequently to be moved ashore again by wave action.

2.3 Other processes of lagoon formation

Although most lagoons have probably originated by the processes outlined above, other evolutionary sequences may be locally important. Along coasts where the land is subsiding, for example, the sea may breach pre-existing barriers and flood low-lying ground. This may include volcanic craters, as in New Zealand. Secondly, some lagoons (e.g. along the eastern coast of Saudi Arabia) are considered to have originated as wave-cut terraces at times during the Pleistocene when sea levels were slightly lower than today; rising water levels and barrier development then formed shallow, flat-bottomed lagoons. Storms may also move masses of shingle from the ends of spits and deposit them on the mainland shore in such a manner as to enclose small lagoons (e.g. some of those at Shingle Street, Suffolk); and a combination of land subsidence, sediment redeposition and river channel changes may form equivalent lagoons in deltaic environments.

One of the more intriguing means by which small lagoons may be formed is illustrated in Fig. 2.3*a*. Here a spit developed (1) which was intersected subsequently by a second one formed by waves arriving from the opposite direction (2). After the enclosure of one lagoon, the original spit continued its outward growth thereby forming a second enclosure (3), and yet again spit-growth alternated to enclose a third lagoon (4). Such paired spits or cuspate forelands frequently occur on capes or headlands in sheltered seas and sometimes join small islands to the mainland (Fig. 2.3*b*).

Fig. 2.3. *a*. Semi-diagrammatic representation of the enclosure of lagoons (*i*, *ii*, *iii*) by the growth of spits on Cape Am'yak, eastern Siberia (see text). After Kaplin, 1959. *b*. The mountainous island of Monte Argentario joined to the Italian mainland by spits, or 'tombolos', with the consequent enclosure of the Laguna di Orbetello.

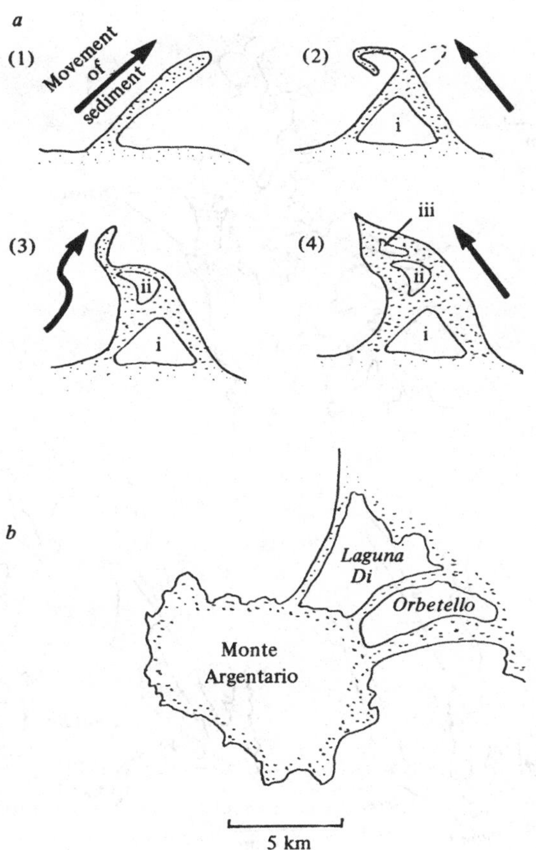

2.4 The morphology of lagoons

Where spits or offshore barriers develop more or less parallel to the original shoreline, elongate lagoons will be formed with their long axes parallel to the coastline (e.g. Fig. 1.3*a*, *c*). However, the rise in sea level mentioned above drowned many river-valley systems and broke into existing freshwater lakes. If these were subsequently isolated by longshore barriers, or by barriers originating offshore and

Fig. 2.4. *a*. The shape of Lake Macquarie, New South Wales, a drowned river valley system subsequently isolated by the development of barriers. *b*. Lagoons near Falmouth, Massachusetts, aligned perpendicularly to the coastline and formed in a glacially highly-modified region.

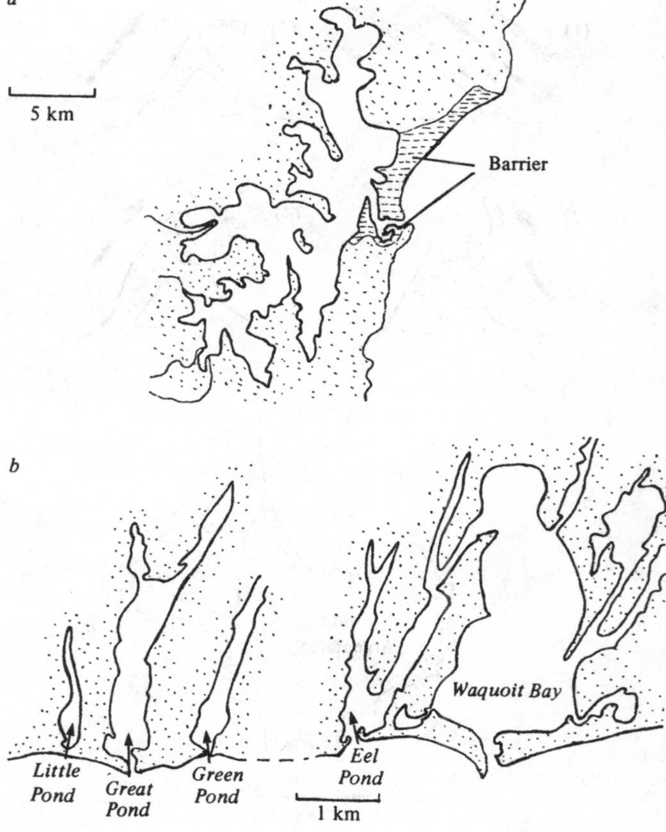

then driven onshore, lagoons of complex outline, including a large element perpendicular to the coastline, will have been formed. Such a history probably accounts for the shapes of Lake Macquarie in New South Wales (Fig. 2.4*a*) and the lagoons of the southern coast of Massachusetts, e.g. those of Cape Cod and Martha's Vineyard (Fig. 2.4*b*). Three thousand years ago Oyster Pond, Massachusetts, for example, was a freshwater lake; then over the next thousand years the sea invaded it, and it remained an embayment of the sea until, only some 200 yr ago, the longshore drift of sand and gravel sealed it from Vineyard Sound (Emery, 1969) to form a lagoon.

Once enclosed, movement of sediment may continue to dominate the history of a lagoon and determine its precise shape. In elongate lagoons, for example, the largest fetch will be along the lagoon's long axis, and hence wind-induced waves and currents will oscillate along its length. These will induce movement of sediment along the shores.

Fig. 2.5. *a*. Diagrammatic representation of lagoon segmentation (see text). From Zenkovich, 1967. *b*. Incipient segmentation of an elongate lagoon in eastern Siberia. From Zenkovich, 1967. *c*. Segmentation of the Kosi lagoon system, Zululand.

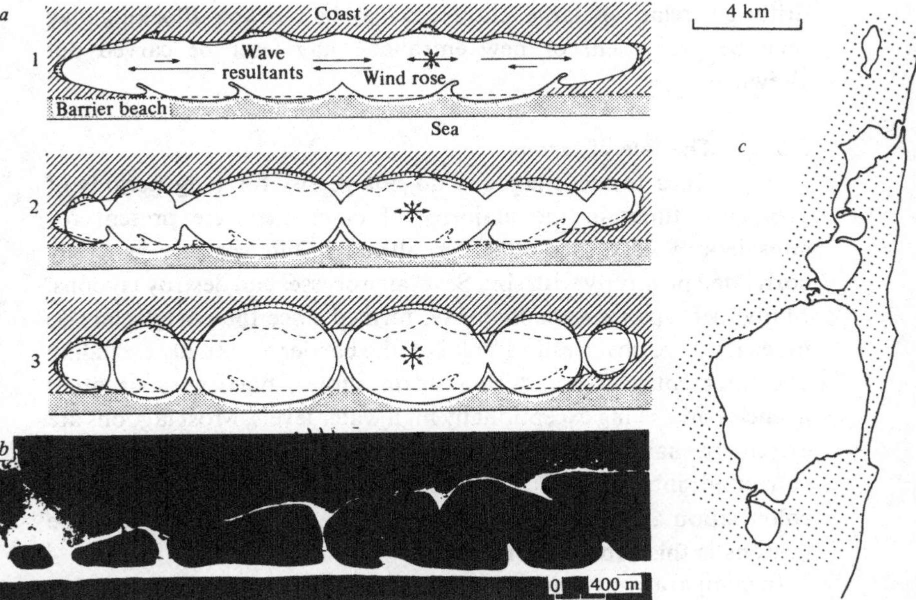

Small spits will form around small irregularities of coastline and will jut out into the water, providing relative shelter in their lee. Weak waves approaching from the other direction can then fill the angle between the base of the spit and the shore with sediment, converting the spit into a cusp. Free of the shelter of the spit/cusp, the predominant waves can build a second spit further along the shore with material eroded from the intervening coast, and so on (Figs. 2.5*a*, *b*, 1.2*b*).

Continuation of this process leads to the segmentation of elongate lagoons into a series of discrete, circular or oval pools, especially in lagoons subject to minimal tidal influence. Such a topography is shown, for example, by the Kosi lagoon system of Zululand, South Africa (Fig. 2.5*c*) (Orme, 1973).

Finally, in this section, we can consider the siting of the lagoon entrance. This is rarely 'arbitrarily' positioned but relates to the pattern of wave-energy distribution along the seaward side of the enclosing barrier, occurring where wave action is weakest, for example as a result of refraction. This notwithstanding, entrances are maintained open only by the high current velocities through them and if current velocities are relatively low at a time when longshore drifting is relatively large, entrances may be blocked. Blocking may then be permanent or new entrances may later be carved out elsewhere.

2.5 The fate of lagoons

Once formed, lagoons do not persist for long periods of geological time; in the majority of cases they are present for considerably less than 1000 yr although life span is generally correlated positively with size. Several processes can destroy lagoons. Many, perhaps even the majority, formed since the postglacial rise in sea level are now freshwater lakes, the barrier eventually excluding sea-water entry except during storm surges, hurricanes or other agencies producing exceptionally high water levels. Most lagoons are extremely shallow (p. 28) and hence they are also relatively easily converted into swamps, marshes and ultimately land by plant colonisation and encroachment (see p. 34). Only in the more saline lagoons is this process insignificant.

In comparatively exposed regions, particularly heavy wave action can break through and erode an enclosing barrier, so that a coastal bay is reformed. Guichen and Rivoli Bays adjacent to the Robe/Beachport lagoons in South Australia (Fig. 2.6) may represent one-time lagoons lost in this manner (Bird, 1970). And finally, a whole barrier may migrate landward, continuing the process described on pp. 7–8, and obliterate the lagoon. Onshore gales may, through wind and wave action, gradually but relentlessly move sediment from the windward face of a barrier, up over its crest, and redeposit it on or behind its leeward face. Thus, for example, several of the sand and shingle barriers around the British Isles, including The Chesil, Scolt Head Island and Orfordness, are slowly moving landward over the lagoons or marshes which have developed in their lee. Eventually, the barriers may become plastered on to the mainland. Such a process will clearly only obliterate elongate lagoons of which the barrier forms one whole side and is equal to, or greater than, the

Fig. 2.6. The Robe/Beachport lagoons, South Australia, showing the re-establishment of two coastal bays by erosion of the enclosing barrier. After Bird, 1970.

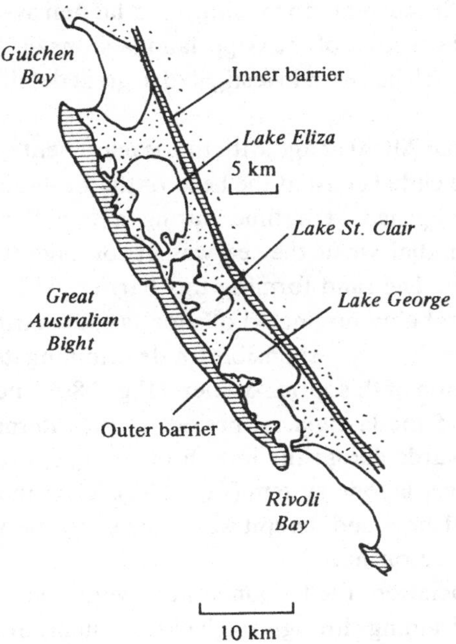

lagoon's length. Of course, relatively minor movements of sediment can obliterate small lagoons situated within a barrier.

Increasingly, man is also affecting the evolution of lagoons by reclamation works, by stabilising the position of entrances and by modification of hydrographic regimes. Many of the small lagoons developed behind shingle ridges in Cornwall, for example, have been reclaimed or converted into freshwater boating or ornamental lakes by damming or sluicing; and the Venetian lagoon has been much altered by human activity. Many Italian lagoons were drained in the effort to combat malaria. It is not all loss, however. Mention of the small lagoons created during reclamation schemes has already been made, and man has, for example, recreated (albeit unwittingly) the Swanpool lagoon in Cornwall by installing an outlet pipe through the completely enclosing shingle ridge. The history of Swanpool has included successive phases of river valley, estuary, lagoon, freshwater lake and now lagoon again (Little, Barnes & Dorey, 1973).

2.6 The Gippsland lagoons as an integrating example

As an example of how all the processes described above may combine to produce the current morphology of a lagoon system, we can examine briefly the history of the Gippsland lagoons in Victoria, Australia (Fig. 2.7) which has been investigated in great detail by Bird (1961, 1966).

The story starts some 70000 yr agc with the development of a sand barrier in front of the cliffed coast at the head of the East Gippsland embayment. This happened at a time during a late Pleistocene interglacial or interstadial when the sea was at, or slightly higher than, its present level. The sand forming this barrier, which partly enclosed a lagoon, probably originated offshore as submarine bars, but longshore drift was largely responsible for determining its shape. Soon after the formation of this 'prior barrier' (Fig. 2.8*a*), and before the fall in sea level of the last glacial period, a spit intermittently extended north-eastwards across the mouth of the bay, eventually enclosing a much larger lagoon system (Fig. 2.8b). The sand of this spit also originated offshore and the spit was widened by the addition of further ridges moved onshore.

During the last glaciation, the lagoon system would have lost its water (probably by draining through the barrier system) and rivers cut their courses through the prior and inner barriers, reducing them to their present dissected condition (Fig. 2.8*c*). On transgression of the sea again, the land was flooded and the lagoonal system reformed (Fig. 2.8*d*); and when the sea achieved its present level, a third barrier, the 'outer barrier', extended both south-westwards and north-eastwards to form the 'Ninety Mile Beach' (Figs. 2.7, 2.8*e*). An elongate lagoon, Lake Reeve, was trapped between the inner and outer barriers. Evidently there were a number of gaps in the outer barrier but these were subsequently closed by longshore drift (and indeed the current entrance channel is an artificial one constructed and maintained by man).

More recently, movements of sediment within the lagoon system have brought about the isolation of Lake Wellington from Lake Victoria, with the result that the former is now freshwater, have built various cusps and deltas in the lagoons, and have segmented parts of Lake Reeve and its eastern equivalent the Cunninghame Arm. Swamp encroachment has obliterated the original lagoon formed

Fig. 2.7. The east Gippsland coast, Victoria, showing the Gippsland lagoons and a larger scale drawing of part of the area (in box). After Bird, 1966. The area in the box is that illustrated in Fig. 2.8.

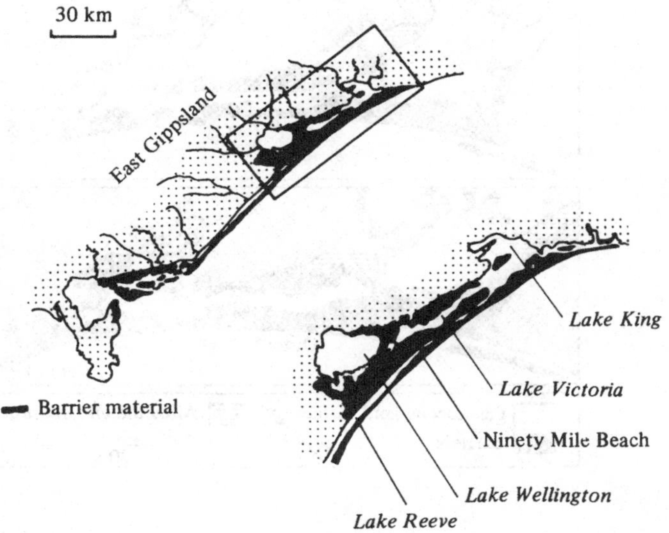

Fig. 2.8. The evolution of the Gippsland lagoons (see text). From Bird, 1966.

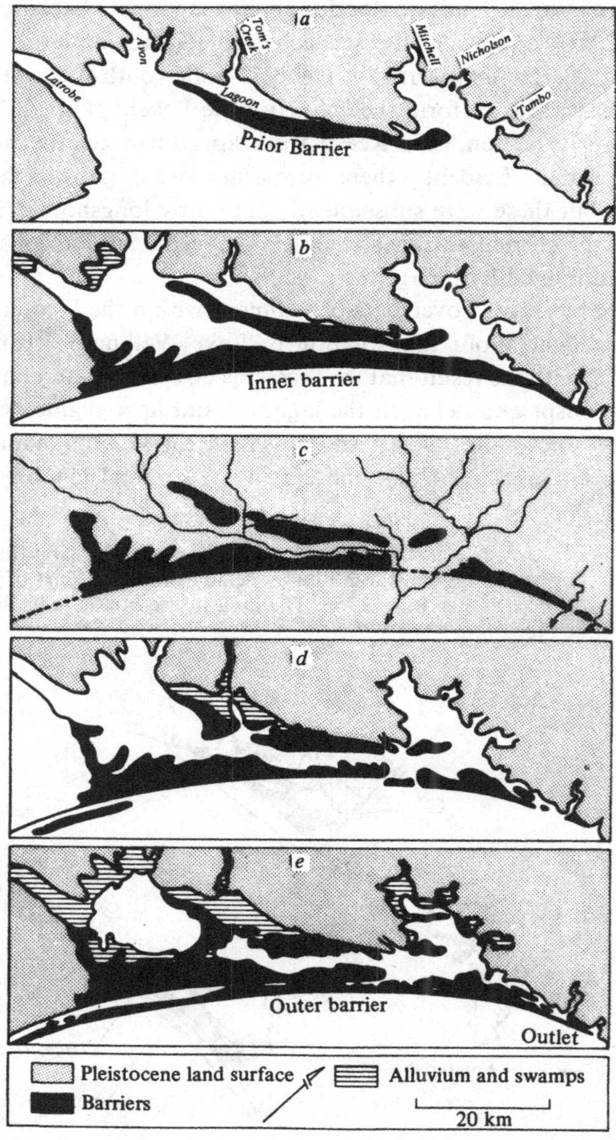

behind the prior barrier and has greatly decreased the area of standing water in Lake Wellington and Lake King. Lastly, it appears that parts of the outer barrier are now experiencing erosion, perhaps as a result of the increasing frequency of coastal storms, with the resultant possibility of the natural appearance of new entrances.

Other recent studies of the evolution of individual lagoons are those of Hill (1975), Hodgkin (1976) and Lankford (1977) (and see the various papers in Castañares & Phleger, 1969, Coates, 1973, and Schwartz, 1973).

3

The lagoonal environment

Any attempt at a concise description of the lagoonal habitat must face the fact that there is not one but many lagoonal environments. There may be as many different lagoonal environments as there are lagoons (Colombo, 1977)! This diversity is reducible, however, to four environmental types, characterised by the salinity of the water, of which any given lagoon may possess from one (as in many small to medium-sized lagoons) to, rather exceptionally, three (as in several of the larger or more elongate systems). These are displayed diagrammatically in Fig. 3.1, and comprise a zone largely influenced by freshwater discharge, an area dominated by water from the adjacent sea, an intermediate region of brackish water, and a hyperhaline environment with salinities in excess of that of normal sea water.

In practice, the region adjacent to the entrance channel and therefore dominated by water from outside the barrier is often marine in nature, but it is not necessarily so. Several of the seas around which lagoons occur are largely or entirely land-locked systems with

Fig. 3.1. An hypothetical lagoon showing the four types of environment which may occur. (Many lagoons may comprise only one or two of these environments.)

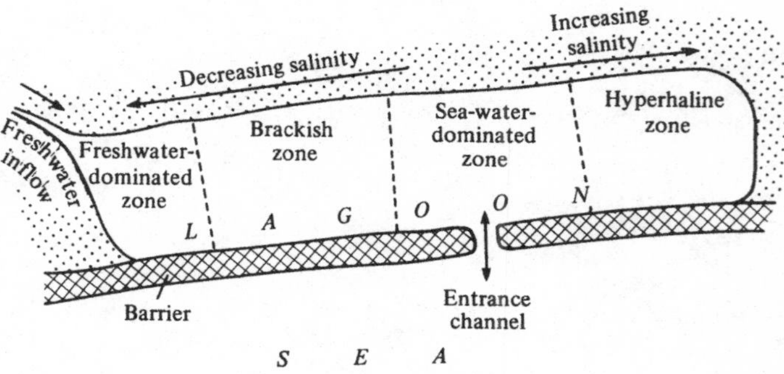

salinities differing markedly from sea water, the southern Baltic Sea (*c.* 6–8‰), the Black Sea (17–18‰), the Azov Sea (11‰) and the Caspian Sea (mostly 12–13‰) for example, and therefore the 'marine zone' may itself be brackish. Further, the occurrence of lagoons within other lagoons complicates the picture, but, nevertheless, the four-habitat-type system still applies to lagoons in general. The lagoons of land-locked seas merely lack the equivalent of the sea-water dominated zone, which is replaced as the site of the entrance channel by the brackish compartment.

Lagoons do, however, exhibit a number of common features regardless of their salinity and so after examining the specific features of the salinity pattern and some of its consequences and correlates, we can consider these general characteristics without detailed reference to precise lagoonal type.

3.1 Salinity

3.1.1 *General considerations*

Because of the relatively small size of the entrance/outlet channel in relation to a lagoon's volume, and in marked contrast to the situation in estuaries (Barnes, 1974), longitudinal salinity gradients in lagoons are stable and do not fluctuate on a semi-diurnal or diurnal basis in relation to tides. They may, however, vary seasonally, especially in areas subject to wet and dry periods of the year, because the controlling factors responsible for the nature of the gradient are volume of freshwater input and volume evaporated from the surface. In the wet season, lagoons may be greatly diluted by freshwater falling in directly as rain and flowing in through river systems, and the freshwater-influenced zone may then extend throughout a lagoon. In contrast, in the dry season when evaporation from the water surface can exceed the volume of freshwater input, salinity may rise to hyperhaline levels. Such a situation is exemplified by the Caimanero Lagoon in western Mexico (Fig. 3.2), where salinity can increase by 2‰ per day in April as a result of the evaporation of 10 mm of water per day from the lagoon's surface (Edwards, 1978). Similarly, northern regions of the hyperhaline Laguna Madre in Texas show a salinity minimum of some 44‰ in March, rising to 70‰ in July (Hedgpeth, 1953), and even in temperate climes, the Comacchio Lagoon in north-eastern Italy shows a salinity increase from 26‰

in March to 48‰ in September (Colombo, 1972). The situation in the Caimanero Lagoon is in fact rendered somewhat complex by human control of the outlet-channel system in the interests of a prawn fishery (see p. 70).

Lagoons are shallow systems (p. 28) and therefore wind-induced mixing often prevents the development of vertical stratification of the water in respect of salinity (again in contrast to many estuaries). Such is not always the case however. Lagoons with a relatively deep basin near to the site of sea-water inflow may possess in this basin a body of high salinity water overlain by less concentrated surface waters. One such lagoon is Swanpool in Cornwall (Dorey, Little & Barnes, 1973). In common with Oyster Pond in Massachusetts and with a number of other land-locked lagoons, sea water only enters Swanpool

Fig. 3.2. Changes in water level and salinity in Caimanero Lagoon, Mexico, in relation to the wet and dry seasons. *a*. Mean monthly water temperature; *b*. mean monthly salinities; and *c*. monthly rainfall figures for the adjacent city of Mazatlan. From Edwards, 1978.

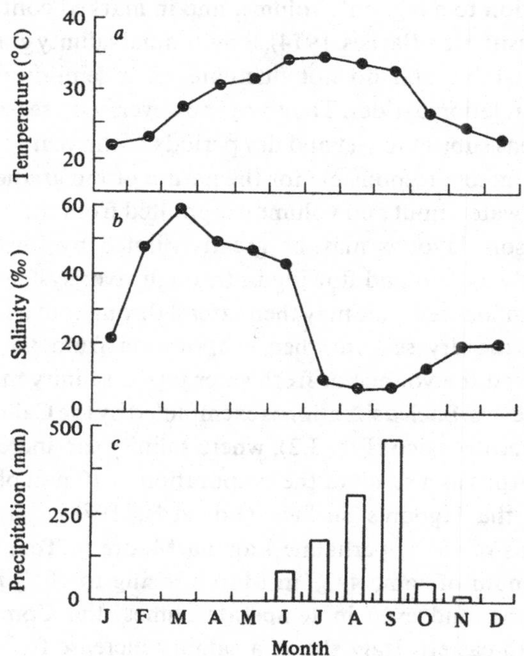

during high water of spring tides. As it flows in, it descends into a 'deep' (2.75 m) basin, being diluted on passage to about half concentration by admixture with surface water. This body of salt water then remains partially separated from the surface water by a sharp halocline (Fig. 3.3). In fact, successive spring tides gradually build up the volume of this 'pool within a pool' causing the halocline to rise nearer and nearer the surface. Salt, however, is gradually entrained into the surface waters, especially by wind action; and during the intervals between spring tides the outflowing surface waters slowly deplete the lagoon's reserves of salt and the halocline descends again. The salinity of the surface waters of the lagoon (Fig. 3.4) therefore varies in a complex manner determined, not by a direct wet and dry seasonal alternation, but by the pattern of sea-water inflow (related to the varying heights of spring tide), by the pattern of freshwater inflow and by the incidence of high winds.

It will have been obvious from this discussion that water level in a lagoon may be intimately connected with changes in salinity. When salinity rises as a result of an excess of evaporation over precipitation, water levels must also fall, and *vice versa*. Thus in the wet season, the Caimanero Lagoon has a surface area of 134 km² and a maximum depth of some 2 m, whereas in the dry season its surface area is only 67 km² and its depth, where water remains, 30 cm or less, a loss of 78% of its volume. Similarly, the Comacchio Lagoon falls from 1.1 m depth in March to 0.6 m depth in September. Each injection of sea

Fig. 3.3. A section through the entrance channel of Swanpool, Cornwall, showing the path taken by inflowing sea water (slopes exaggerated by a factor of four). The figures on the diagram are in ‰ salinity. A vertical salinity profile, showing a marked halocline, is given at the right (data of 4 July 1968). After Dorey *et al.*, 1973.

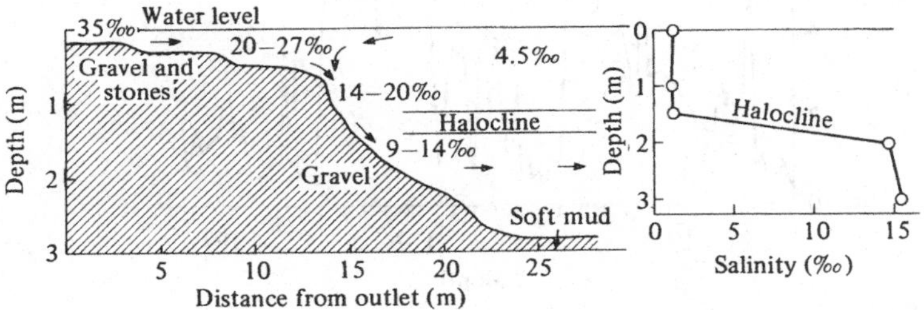

water into Swanpool, besides increasing the mean salinity, both raises the level of the lagoon and, during periods of low tide, increases the outflow from it, although there is a net gain in pool height during periods of sea-water inflow (Fig. 3.5), the opposite correlation between salinity and water level to that seen in the lagoons considered earlier. Indeed, the two environmental variables need not be related at all. Small lagoons within shingle barrier systems, for example those behind the shingle ridge at Porlock, Somerset, fill with sea water by percolation through the shingle during periods of high tide in the adjacent bay, and empty again during low tide. The fauna, at Porlock dominated by the isopod *Sphaeroma*, spends the periods when the lagoons are without water in moist micro-habitats, underneath pebbles, etc. The salinity of the water in such lagoons remains constant during the inflow/outflow cycle. In a sense, the Swanpool and Porlock examples are atypical in their dependence on a tidal regime and – and the two factors are related – in being isolated behind shingle barriers, although these circumstances are relatively common in Britain (Barnes & Heath, 1980). More generally throughout the world, the inverse relationship between salinity and water level would hold, even to the extent of converting what in the wet

Fig. 3.4. Variation in surface salinity of Swanpool, January 1975–May 1977. After Barnes, Williams, Little & Dorey, 1979.

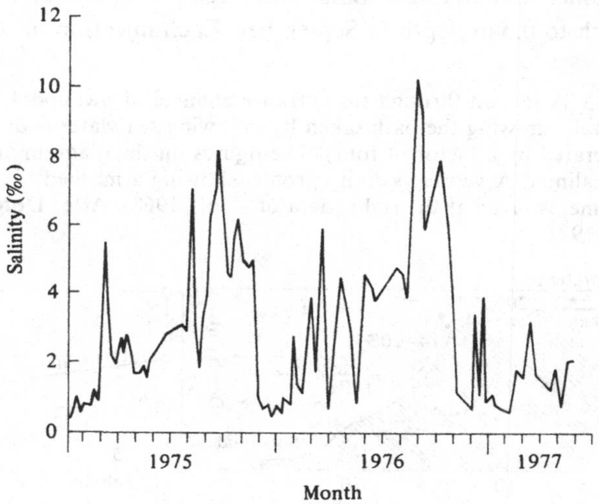

season is an estuary into a series of isolated hyperhaline lagoons during the dry season, as, for example, is seen in the St Lucia system in South Africa.

3.1.2 *Salinity as a factor differentiating lagoonal environments*

The region influenced by the adjacent sea. The region adjacent to the entrance channel in Fig. 3.1 is, as might be expected, that most similar to the adjacent sea. If the adjacent sea is tidal, then the sea-water influenced zone will also be tidal although, because of the bottle-neck formed by the narrow entrance, fluctuations in water level are likely to be less marked than, and delayed with respect to, those in the parent water mass. Tidal fluctuations in this zone of the Dawhat as Sayh Lagoons in the Arabian Gulf, for example, are a maximum of 0.6 m compared to 1.6 m in the Gulf, and they occur

Fig. 3.5. Exchange of materials through the entrance channel of Swanpool during a series of high spring tides. *a.* Water level of lagoon surface. *b.* Cumulative curve of salt transfer (expressed as volume of sea water). *c.* Cumulative curve of transfer of water. *d.* Salinity of water in the channel. After Dorey *et al.*, 1973.

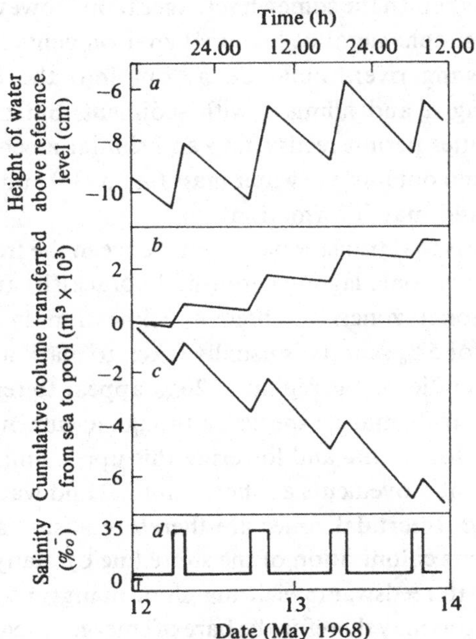

after a delay of 2–3 h (Jones, Price & Hughes, 1978). Because of the buffering effect of the adjacent sea, annual salinity fluctuations are least in this zone and so, notwithstanding possible tidal fluctuations, is variation in water level on an annual wet/dry season basis. These factors combine to permit colonisation of the intertidal zone by salt-marsh or mangrove-swamp vegetation (see p. 34), and lagoons comprising only this zone closely approximate to highly sheltered, marine bays or inlets, although the greater degree of enclosure may have important consequences for their productivity (see pp. 39, 49–52).

Regions hypohaline to the adjacent sea. At one of the extremes of the potential lagoonal salinity gradient, the freshwater-influenced zone, is the most stable of lagoonal environments, at least in the short term. Except under dry-season conditions, if such occur, salinity fluctuation is very small and water level varies little. Even at times of greater than average freshwater flow, conditions typical of this zone may merely extend further into the lagoon system and the zone itself remain virtually unchanged, although current velocities – and therefore turbidity – will be higher. In the somewhat longer term, however, this zone can be the most ephemeral of lagoonal environments because not only can inflowing rivers build deltas out into the lagoon, gradually segmenting it and filling it with sediment, but also the prevailing low salinities permit reedswamp and similar assemblages of vegetation to extend out into the water mass (see p. 34) and reduce its area by swamp and marsh formation.

The brackish zone is that transitional region between the fresh and marine areas (although some lagoons are wholly brackish) and, like most other transitional zones, its precise delineation is almost impossible. A value of 5‰ salinity is usually taken to mark its lower boundary; whilst salinities in the region of 20‰ appear to represent a boundary to many truly marine species, although arguments have been advanced both for raising and lowering this upper limit of the brackish zone. Current movements are here minimal and water level may hardly fluctuate. Intertidal zones are therefore absent and the brackish water inhibits colonisation of the shore line by many of the species associated with reedswamp, salt-marsh or mangrove-swamp formations: the margins may therefore be bare of emergent vegetation (but see p. 35).

Regions hyperhaline to the adjacent sea. Some lagoons, or regions of lagoon systems, are permanently hyperhaline, even to the extent that salt may crystallise out around the shores and on the bed, as in the Kara-Bugas lagoon off the eastern Caspian Sea where salinities can achieve 260‰ (a twenty-fold concentration of the inflowing water). The salt (particularly sodium sulphate, Glauber salt), which precipitates out in winter, supports a local industry. In these regions of almost permanent excess of evaporation over precipitation, current flow in lagoons is reversed. Normally, lagoons show a net discharge of water to the adjacent sea, but under these conditions water evaporated from the surface draws water from the sea into the body of the lagoon and this too is evaporated as it flows down the water-level gradient. The level of the Kara-Bugas Lagoon is some 3 m lower than

Fig. 3.6. Gradient of salinity along two hyperhaline lagoons in relation to the site of the entrances. *a*. The Laguna Madre, Texas (average values for 1946–8). *b*. The Sivash, Crimea (average values for 1935–6). After Hedgpeth, 1957.

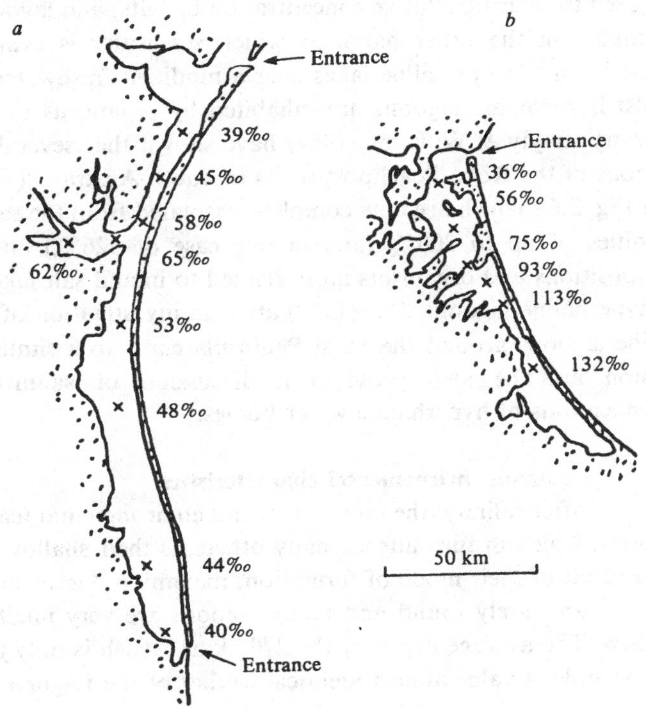

that of the Caspian Sea, for example, and a volume of 12–25 km^3 is drawn in from that Sea per year. Regions furthest from the entrance channel are therefore the most hyperhaline (Fig. 3.6).

Hyperhaline conditions (which normally cover the range 45– > 130‰ salinity) are even more inimical to fringing vegetation communities than are brackish ones, and therefore shores may also be unvegetated. In extreme cases, like that of the Kara-Bugas Lagoon, animal life may be impossible, although algal numbers may be vast; 21 × 10^6 micro-algae, flagellates, etc., per cm^3 of Kara-Bugas water, for example. Permanently hyperhaline lagoons will clearly lack the freshwater-influenced and brackish environments. To evolve into a freshwater lake (p. 14) is not the only possible fate of lagoons completely isolated from the sea. Isolated hyperhaline lagoons may become even more hyperhaline and some evolve eventually into hypersaline lakes. The distinction between hyper*h*aline and hyper-*s*aline is based on the faunal, floral and chemical differences between inland salt lakes on the one hand (for which saline and hypersaline are used to indicate relative concentration of salts) and lagoons, for example, on the other hand, in which sea water is evaporated (hyperhaline). Hypersaline lakes bear a modified freshwater biota, whilst hyperhaline lagoons are inhabited by organisms of marine ancestry. Bayly & Williams (1966) have shown that several of the lagoons of the Robe/Beachport series in South Australia (see p. 15 and Fig. 2.6), which are now completely isolated from the sea, have salinities of up to 200‰ (and in one case of 326‰) and ionic compositions and organisms more related to inland salt lakes than to hyperhaline lagoons. Por (1972) after an investigation of hyperhaline lagoons around the Sinai Peninsula came to a similar conclusion and he also provides a discussion of salinity-based classifications of hyperhaline water bodies.

3.2 Common environmental characteristics

After salinity, the most important environmental feature of lagoons, which in turn affects many others, is their shallow depth. As a result of their mode of formation, maximum depths in excess of 10 m are rarely found and many lagoons are very much more shallow. The average depth of the 2700 km^2 Sivash is only 0.75 m, for example, a value almost identical to that of the Laguna Madre (maximum depths are both in the region of 3 m, although a longitudinal channel in the Laguna Madre was artificially deepened in 1949 to 4 m by the construction of the 'Intracoastal Waterway'); and the depth of the 260 km-long lagoon system stretching from Cotonou to the Niger Delta, and centred on Lagos, Nigeria, is little more than 1 m (Webb, 1958; Hill & Webb, 1958). In general, lagoons formed from drowned and subsequently enclosed river valleys and freshwater lakes are deepest and may have deep basins within them, sometimes exceeding 20 m. Several consequences of their shallow depth have already been referred to and others will be considered below.

Similarly to estuaries (Barnes, 1974), and basically for the same reasons, lagoons are floored by soft sediments; but, in contrast to estuaries, large areas of intertidal flats do not occur because of the stability of the water level, although lagoons which shrink in size during a dry season (e.g. the Mexican lagoon referred to on p. 23) do expose seasonally somewhat equivalent stretches of sand or mud. As most enclosing barriers are essentially sandy, at least one shoreline will be formed by comparatively coarse-grained sands and, in drowned river-valley systems, etc., erosion by wave action of landward shorelines may give rise to gravelly or pebbly lagoonal coasts. The shores of lagoons within shingle barriers will naturally be particularly coarse grained. The bed of the lagoon itself, however, is usually formed of soft muds because of the deposition of fine sediments in the relatively sheltered prevailing conditions. These are mainly contributed by the inflowing rivers. Particle size of the sediments may therefore be particularly small in the freshwater-dominated and brackish zones, and increase towards the entrance where more vigorous water movement keeps silts in suspension. Because of their shallowness, wind action can penetrate to the bed of most lagoons and fine sediments, which are relatively easily resuspended, render the water turbid during periods of high wind. The rate of sedimentation, which is increased by submerged and emergent vegetation, is then a compromise between the rate of input of materials by the river system and the rate of resuspension and outflow; usually there is a net accumulation of sediment. Storms may also deposit fans of sediment eroded from the barrier in the main body of the lagoon. As a result of various chemical and physical processes, lagoonal substrata are often composed of uncompacted, fine, 'fluffy' floccules, rather than the stiff muds characterising many estuaries.

Lagoons are frequently nutrient-rich (Mee, 1978), both as a result of the input of nutrients by rivers and through the effectiveness of recycling between sediment and water mass (another consequence of their shallowness); phosphate levels in the surface waters of L'Étang Bages-Sigean, France, for example, achieve 50 μg P-PO_4 per litre whilst those in the interstitial waters of the sediments attain twenty times this amount (Fiala, 1973). They are therefore often highly productive (pp. 49–52) and there may be a large input of organic matter into the surface sediments, which will require large quantities of oxygen for its decomposition. Because of the well-mixed nature of the water, oxygen may reach the sediments in sufficient quantity to meet this need, although the sediments themselves may be anoxic only just below the surface; but in lagoons with a very large productivity or with deep basins and/or with haloclines, anoxia of the bottom waters is common (Fig. 3.7). Biological activity, other than of anaerobes, is then confined to the surface water layers and to the relatively shallow areas overlain by those waters.

Evaporation of water from the surface of hyperhaline lagoons has two effects relevant to these processes. Nutrients may be concentrated to very high levels, further raising potential plant productivity (although such high nutrient status is not always found) and, because

Fig. 3.7. Vertical oxygen profile in Swanpool taken at the same time as the salinity profile depicted in Fig. 3.3, showing the pronounced oxycline and the anaerobic bottom water. After Dorey *et al.*, 1973.

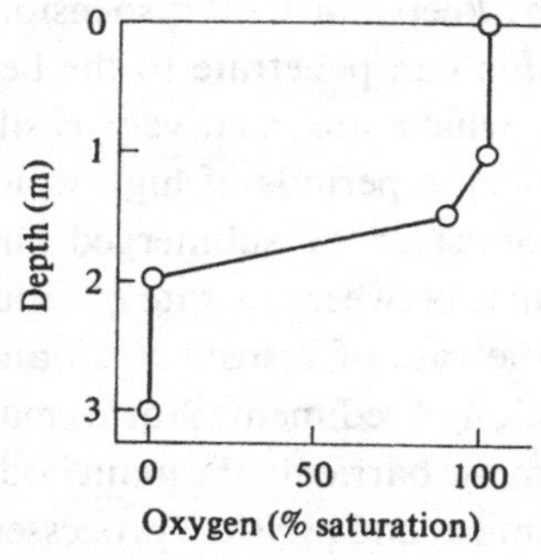

of the increased density of the evaporated surface waters, water continually sinks from the surface to the bed taking with it oxygen. Solubility of oxygen in water decreases, however, with both increase in salinity and increase in temperature.

Most of the other characteristics of the lagoonal environment are

Fig. 3.8. *a*. Daily variation in mean dissolved oxygen in Caimanero Lagoon, Mexico, at a time (September, 1974) when salinity was 10‰ and temperature was 29 °C (vertical bars indicate the range). From Edwards, 1978. *b*. Annual variation in some physico-chemical conditions in the Comacchio Lagoon, northern Italy. The ranges are: temperature, 8.5–29.0 °C; O_2 saturation, 50–150%; pH, 7.7–9.5; nitrite, 0.01–5.1 μg-at N per litre; nitrate, 0.05–9.3 μg-at N per litre; silicate, 0.1–4.1 μg-at Si per litre. After Colombo, 1977.

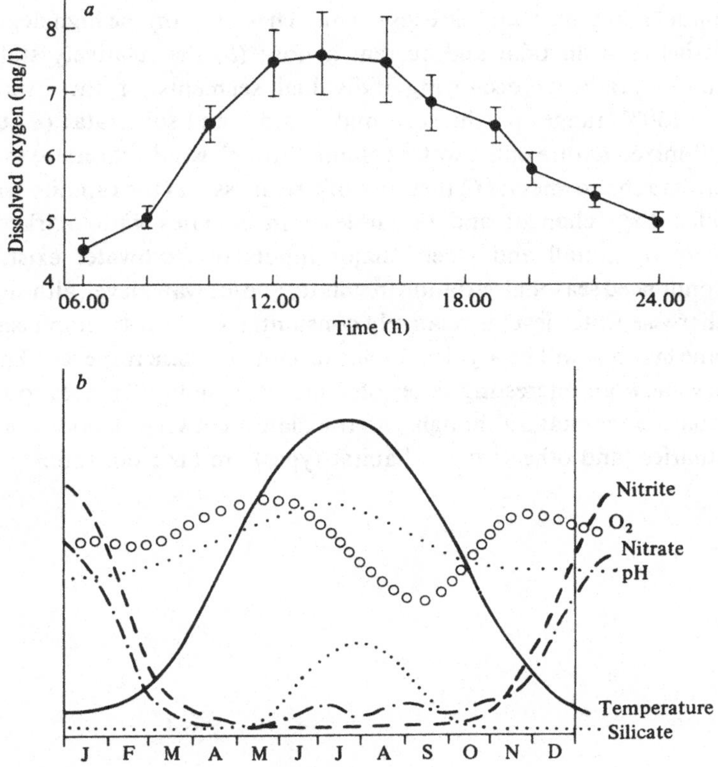

those common to all shallow, pond-like bodies of water (see Barnes & Mann, 1980). Temperature variation will reflect that of the overlying air, for example, and very high temperatures may be recorded in the shallows; pH is more variable than in sea water; in highish latitudes, relatively fresh lagoons may freeze over in winter; and so on. Many physico-chemical variables will, as in most aquatic habitats, vary during the day and during the year as a result of biological activity, particularly the respiratory/photosynthetic gases (and through them pH) and the various plant nutrients (nitrate, phosphate, silicate, etc.). Examples of this biologically-induced variation are given in Fig. 3.8.

Although these factors are, of course, of great significance to lagoonal organisms, and just as important as variables more specific to lagoons, they are not the ones which give lagoons their distinctive character amongst aquatic ecosystems. These are: (*a*) the high degree of shelter from tidal and current action; (*b*) the relatively stable salinity gradients occupying individual segments of the overall 0–> 130‰ range; (*c*) the soft mud and/or sand substrata; (*d*) the well-mixed nature of the water column through wind action; (*e*) their extreme shallowness; (*f*) their organic richness; (*g*) the rapidity with which they change; and (*h*) at least in climates with markedly seasonal rainfall and where major inputs of freshwater exist, a pronounced seasonal variation in salinity and/or water level, although otherwise water level is relatively constant. (Needless-to-emphasise, some lagoons will be atypical in one or more of these respects.) Thus they show an interesting series of similarities and differences to the estuarine habitat, although intermediates between lagoons and estuaries (and other aquatic habitat-types) are far from rare.

4

Lagoonal ecology

4.1 Introduction

Brackish-water environments have long fascinated biologists, mainly, it would appear, because of the physiological adaptations which their organisms have evolved and because of their potential for studying the distribution of organisms along environmental gradients (particularly that of salinity). Many lists of local fauna and flora have been published, often with corresponding listings of salinity-tolerance ranges as evidenced by field distributions; but, until comparatively recently, very little more was known of lagoonal ecology and much still remains to be investigated.

Species lists obviously are important and they often form an essential starting point for an ecological study; but, when one considers that lagoons range from the almost freshwater to the hyperhaline and occur from the Siberian arctic to the equator and from the inland seas of the Eurasian land-mass to the Polynesian islands in mid-Pacific, the range and diversity of species occurring in them in total is enormous, and the majority of names in a composite list would be completely unknown to anyone familiar with only one part of the world. Hence I will not attempt to summarise this vast literature or to provide lists of typical lagoonal species, but, rather, will confine myself to the general types of organism typically occurring, especially in relation to those to be found in the parallel estuarine habitat, and to what is known of the interactions of these species. Lists of lagoonal species may be consulted in: Zenkevitch (1963), Muus (1967) and Remane & Schlieper (1971) (northern Eurasia); Marchesoni (1954), Jacques *et al.* (1975) and the references cited by Sacchi (1979) (Mediterranean); Broekhuysen & Taylor (1959) and Boltt & Allanson (1975) (South Africa); Bayly (1967) (Australia); Hedgpeth (1967) (Gulf of Mexico); Holm (1978) (Florida); and in many of the other individual papers listed on pp. 95–102.

Because many summaries of freshwater and coastal marine ecology are available, I will also largely ignore those lagoons, or those parts of lagoon systems, which differ but little from freshwater lakes or from the adjacent sea.

4.2 Lagoonal organisms

4.2.1 *Fringing macroflora*

The semi-aquatic emergent macrophytes of lagoons belong to the same three ecological groupings as characterise estuaries: mangrove-swamp, salt-marsh and reedswamp. Mangroves dominate the sea-water influenced zone in the tropics and various species also extend well into the brackish and hyperhaline zones. *Avicennia* and *Lumnitzera*, for example, are said to be able to withstand salinities of up to 90‰, albeit in a dwarfed condition; whilst *Rhizophora* can tolerate a range of 12–55‰. Tolerance of low salinities is generally more common than that of hyperhaline environments, however, although few mangroves are found in waters of less than 10‰ salinity. There are some indications that species of *Avicennia* are more characteristic of lagoons with fluctuating water levels, and species of *Rhizophora* and the appropriately-named *Laguncularia* of systems with more constant depths of water.

In temperate and relatively dry regions, salt-marsh or salt-desert vegetation shows a broadly similar distribution to the mangroves of hot and wet climes; although salt-marshes are more restricted to zones of lagoons with a regular tidal regime. Nevertheless, species otherwise characteristic of salt-marshes may extend in a thin line along the shore (e.g. species of *Salicornia*, *Arthrocnemum*, *Suaeda* and *Kochia*, all Chenopodiaceae), and cord-grass, *Spartina*, which is more tolerant of submersion than other salt-marsh plants, can colonise the shallows, even extending into the freshwater-influenced zone. Several of the Chenopodiaceae can tolerate hyperhaline (and hypersaline) conditions and occur as a salt-desert flora around hyperhaline lagoons.

Rather more characteristically lagoonal in temperate regions are variants of reedswamp. The common reed, *Phragmites australis* (= *communis*), is normally associated with freshwater habitats but it can also withstand stable salinities in excess of 20‰, although it is not often tolerant of wide fluctuations in water level. It may therefore colonise the whole of the brackish and freshwater zones of temperate lagoons; it clothes, for example, the margins of the Black Water Lagoon near Beaulieu, Hampshire, which has a salinity of 18–19‰. Associated with the reed may be other tall grasses and sedges which can achieve local dominance, especially reed-maces, *Typha*, which can tolerate up to 10–15‰, and several species of the sedge *Scirpus*, some of which can survive in full-strength sea water.

4.2.2 *Submerged macroflora*

The quiet, shallow, nutrient-rich waters of lagoons form an idea habitat for submerged macrophytes which can attach themselves to the soft sediments by rhizomes, or rhizome-like structures, and extend long linear or strap-shaped leaves into the water. These grass-like plants are often able to survive and reproduce under conditions of total submersion. Six related groups of monocotyledons occur. The eel- and turtle-grasses (Zosteraceae and Hydrocharitaceae) are typical of the more marine regions of lagoons, and they are replaced in brackish zones by grass-like pondweeds (Potamogetonaceae, e.g. *Potamogeton pectinatus*), horned pondweed (Zannichelliaceae), the tasselweeds (Ruppiaceae) and the naiads (Najadaceae). Several of these brackish waterweeds are widely distributed. *Potamogeton pectinatus*, for example, dominates lagoons in north-west Europe and in South Africa, and *Ruppia maritima* occurs in lagoons in both north-west Europe and Mexico, and they can extend into marine conditions in the absence of the sea-grasses. Both the submerged and emergent macrophytes bear abundant growths of smaller epiphytic algae.

Co-dominant with the pond- and tasselweeds are often found species of aberrant green algae, the Charophytes. These, frequently lime-encrusted, algae with their whorls of 'leaves' are particularly characteristic of clear waters; they form the pioneer colonists of clean sandy areas in the Comacchio Lagoon, Italy, for example, and create a dense ground carpet. Other green algae, particularly species of *Enteromorpha*, occur somewhat sporadically, as they do in estuaries, in areas of freshwater inflow and in those of particularly high nutrient concentration.

The submerged macrophytes do not tolerate hyperhaline conditions well. One exception is the sea-grass *Halodule*, which occurs, for example, in the Laguna Madre and can withstand there salinities in excess of 50‰. Another is the genus *Halophila*, which in Saudi Arabia can tolerate at least 60‰.

4.2.3 *Plankton and micro-organisms*

Relatively little work has been devoted to the nature of the smaller organisms in lagoons, although there is ample evidence that their biomass and productivity are often high. Lagoonal phytoplankton, especially, is both rich and diverse and, because of the ability of wind action to penetrate to the bed, is frequently enriched by benthic algae put into suspension. Although, as in other aquatic systems, diatoms are important components of the phytoplankton (see, e.g., Gilmartin & Revelante, 1978), typically lagoons show a comparatively large contribution by dinoflagellates, chlorophytes, cryptophytes and other micro-flagellates. Toxic blooms of dino- and other flagellate species have not infrequently been recorded (see, e.g., Comin & Ferrer, 1978).

Although the phytoplankton is usually abundant, the same does not appear to be true of zooplankton, whose distribution and abundance can only be termed irregular and poorly understood. Some lagoons do certainly teem with zooplanktonic organisms, although often of a very limited range of species, and, for example, lagoonal water may appear cloudy with ostracods or with copepods such as *Eurytemora* or *Acartia*; in other lagoons, however, zooplankton is scarce or effectively absent and its 'place' is taken by small nektonic organisms such as mysids. In addition to such permanently planktonic species as may be present, several benthic species ascend into the plankton at night or are suspended by wind action, and some benthic animals have planktonic larval stages. Thus crustaceans, such as *Gammarus*, cumaceans and mysids, and insects, such as dipteran larvae (especially of chironomids), water-bugs and beetles, may be planktonic during part of the day; harpacticoid copepods may be injected in large numbers into the water during rough weather; and some polychaetes and molluscs may spend brief parts of their lives as free-swimming larvae. Planktonic ciliates are numerous in some lagoons but have rarely been investigated.

With the exception of the visually-obvious surface mats or felts of diatoms and blue-green algae, together with their included micro-flagellates, ciliates, etc., benthic micro-organisms are even more poorly known; although from the scanty evidence available it seems likely that lagoons are very similar to estuaries in this regard. Whenever examined, except in anoxic regions, large numbers of diatoms, blue-green algae, ciliates, micro-turbellarians, nematodes, harpacticoid copepods, small annelids, halacarid mites and other meiofaunal animal groups have been observed, and relatively large bacterial counts have been found in anoxic zones, as elsewhere within lagoons. Several of the meiofaunal animals have very wide geographical ranges; the small, tentacle-less hydra, *Protohydra leuckarti*, for example, is known from Europe, North America, Africa and eastern Asia. Some, however, are only known from a few scattered localities, although it is likely that as more effort is devoted to the biota of lagoonal sediments so more records will be forthcoming. At the moment, the minute anemone *Nematostella vectensis* is known only from lagoons in East Anglia, the Isle of Wight, Woods Hole (Massachusetts), Bay Farm Island (California) and Nova Scotia (see Williams, 1973).

4.2.4 *Macrofauna*

It may have been noted from the sections above that the brackish zone of lagoons supports a fauna and flora of mixed ancestry, with the majority of groups being generally characteristic of marine environments but with a large contingent from freshwater habitats. In this respect they clearly resemble estuaries, but the extent of faunal 'mixing' has been carried further in many lagoons; so much so that insects in particular may dominate the fauna of those lagoons with salinities of less than about 10‰ (and exceptionally up to 15‰) and of those with extreme hyperhaline conditions. The absence of insects from the middle part of the salinity range is probably a result of competition with the essentially marine species: salinities of this order are withstood in inland saline waters. Dipteran larvae, especially those of chironomid midges, are found throughout the lagoonal salinity spectrum, however, and in some lagoons, e.g. the Varna Liman in Bulgaria, they are dominant or co-dominant.

In relatively dilute brackish zones, dipteran larvae are joined by many species of hemipteran bugs (e.g. water-boatmen of the genera *Notonecta*, *Corixa*, *Sigara*, etc.), several beetles, and, more rarely, the larvae of dragonflies and caddis flies, amongst others. Other predominantly freshwater species which can occur in these zones, at least in lagoons with stable salinity regimes, are pulmonate snails, flatworms and leeches. The reasons why such species do not colonise the equivalent zones in estuaries probably relate to the more marked fluctuations in salinity and water level characterising estuaries and to the stronger water movements associated with tidal regimes. One further group normally thought of as being freshwater, the oligochaetes, do in fact occur in soft, often intertidal, marine sediments as well, and so their abundance in lagoonal silts requires less specific explanation; the species concerned mainly belong to one family, the Tubificidae.

Besides these freshwater imports, the benthic macrofauna is dominated by three groups: annelids, crustaceans (especially isopods, amphipods and decapods) and molluscs (gastropods and bivalves). This applies regardless of geographical region although, naturally, the relative importance of the three groups varies from lagoon to lagoon. The majority of these animals are related, often very closely, to species characterising estuaries and the sea (pp. 54–64); indeed many of the species can also be found in such habitats. All are species burrowing in, or living on, soft sediments, or associated with the macrophytes. Several genera have very wide distributions; hydrobiid snails, for example, are characteristic lagoonal animals in the temperate regions of both hemispheres (in the tropics they are replaced by equivalent gastropods, e.g. of the genus *Cerithidea*), and the amphipods *Corophium*, *Gammarus* and *Grandidierella* occur in both temperate and tropical lagoons. Similarly, nereid and spionid polychaetes and several of the isopods are almost ubiquitous; although, in contrast, crabs form an important component of the fauna only in tropical and warm temperate zones.

Although mysids may be important members of the nektonic lagoonal macrofauna, the two groups which dominate most lagoons are prawns and fish. Both may occur in such abundance, even if only seasonally, as to support fisheries (see pp. 69–70). The prawns are mainly species of *Palaemonetes* and, in the tropics, of penaeids (especially *Penaeus*, but also of *Metapenaeus*). The penaeids typically breed in the adjacent sea and enter lagoons as postlarvae, migrating out again when they are from half to fully grown; *Palaemonetes*, however, are either permanently lagoonal or else migrate in to breed.

The fish fauna of many lagoons is large and diverse; sixteen species are described as 'the most common' in the 134 km^2 Caimanero Lagoon, for example, and even in the hyperhaline Sivash, twenty species regularly occur (together with an additional thirty 'occasionals'). Similar numbers frequent the Laguna Madre. The precise species vary on a regional basis, although grey mullet (*Mugil*) are especially widespread and characteristic; other typical fish are anchovies, atherinas, gobies and flatfish, and, in northern European lagoons, eels and sticklebacks. Adult fish enter lagoons to feed, with the result, for example, that 'fish fed in the Sivash grow faster and fatter' (Zenkevitch, 1963) and juveniles, especially of the atherinas, mullet, etc., enter for the same purpose (see, e.g., Warburton, 1978). Chapman (1971) estimates that the fish using the Gulf-of-Mexico lagoons during part of their life cycle, menhaden, spot, Atlantic croaker, striped mullet, and the black and red drums, contributed 31% of the commercial fish catches of the USA in 1968 and earned 27% of its revenue.

4.3 The food web

Lagoonal food webs are broadly similar to those of estuaries, but they differ in two important respects. The contributions made by phytoplankton and submerged macrophyte production are considerably more important in lagoons, and both these and other sources of food are consumed within the system to a much greater extent. There is, therefore, less of a through-put of nutrients and fixed organic matter, a result of the more closed nature of lagoons and of the unimportance of tidal fluxes, although migratory nektonic species do, as we have seen, still enter lagoons to feed and having fed, move out again (see pp. 45–49), and some lagoons may still produce an export to the adjacent sea. The Izembek Lagoon, Alaska, for example, exports large quantities of floating eel-grass leaves to the Bering Sea (Barsdate, Nebert & McRoy, 1974).

4.3.1 *Food sources*

Basically, potential food sources in lagoons are: phytoplankton, benthic and epiphytic algae, and detritus derived from the macrophytes. As in other aquatic ecosystems, very little direct consumption of living macrophytes occurs (except by echinoids and some vertebrates). Phytoplankton is consumed *in situ* by zooplanktonic copepods and by the smaller nekton, especially mysids; but the relative paucity of lagoonal zooplankton may result in incomplete utilisation of this food source, so that considerable quantities of phytoplankton are available to benthic organisms (Fig. 4.1) in the form of sedimenting algal cells or as detritus.

The major source of detritus, however, is the submerged and, often to a lesser extent, the fringing vegetation. Decomposition of a few macrophytes has been studied in lagoonal environments, although most research in this field has been carried out in other habitats. *Potamogeton pectinatus* has been shown to decay very rapidly in the South African Swartvlei Lagoon, for example (Howard-Williams & Davies, 1979). Under environmental conditions of 15–26 °C temper-

Fig. 4.1. The relationship between net phytoplankton production and zooplankton consumption in the Cochin Backwater, Kerala, India, showing a large ungrazed 'surplus' of phytoplankton. After Qasim, 1970. In this case, the surplus is attributable to the paucity of zooplankton.

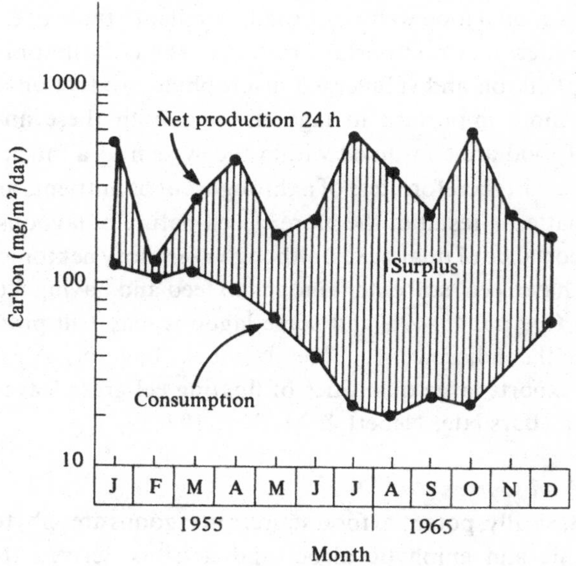

Fig. 4.2. Decomposition of *Potamogeton pectinatus* in Swartvlei Lagoon, South Africa. From Howard-Williams & Davies, 1979. Day 0 = 19 July 1976. *a*. Decay curve. *b*. Left hand series, percentage of original total quantity of various components remaining at various intervals up to 158 days. Right hand series, increase in amount of nitrogen and phosphorus relative to total dry weight over the same time period, showing the effects of the presence of associated microbes growing on the *Potamogeton* substrate.

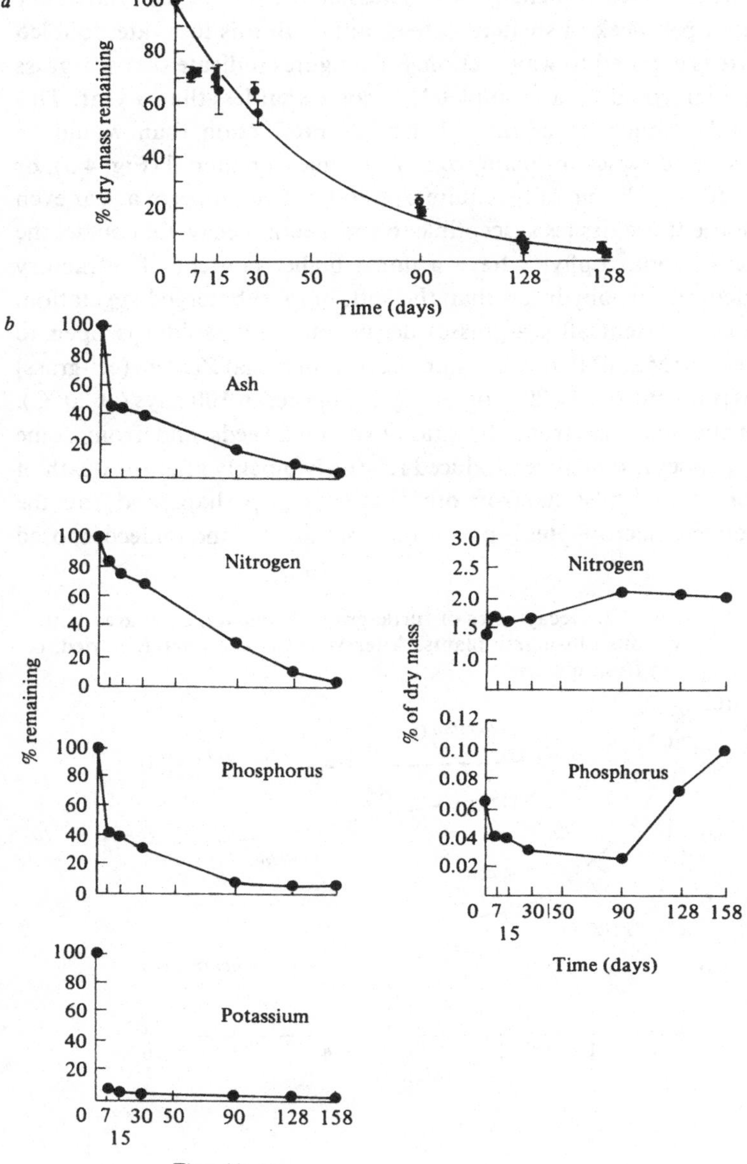

ature and 5–11‰ salinity, most nutrients were released from the plant in the first week of decay (almost all the potassium and 60% of the phosphorus) and complete decay had been achieved by 160 days (Fig. 4.2). Sea-grasses appear to decay more slowly: Zieman (1968) (quoted in Wood, Odum & Zieman, 1969) recorded that dead and detached leaves of turtle-grass, *Thalassia*, lost 10% of their initial dry weight per week in sheltered areas, although this loss rate doubled in areas exposed to wave action. Such figures indicate that sea-grass material would decay completely in somewhat less than a year. This is still a much faster rate of detritus production than would be considered typical of mangrove or salt-marsh material (Fig. 4.3), or of reeds which generally require a period well in excess of a year even to loose those tissues susceptible to short-term decay. Of course, the emergent macrophytes have a much higher content of refractory structural carbohydrates than the soft, limp, submerged vegetation. To what extent all sea-grasses decay relatively rapidly is open to question; Mann (1976), for example, considers that *Zostera* (eel-grass) looses no more than 20% of its organic matter in 100 days (at 20 °C). Detrital material from the grass-like pondweeds and from some sea-grasses is therefore produced relatively quickly after the death of their leaves, whilst that from other sea-grasses, perhaps, and from the emergent macrophytes is produced more slowly; and undecomposed

Fig. 4.3. Decay rates of turtle-grass, *Thalassia*, compared with various salt-marsh plants. After Wood *et al.*, 1969. (d) dried, or (w) fresh materials.

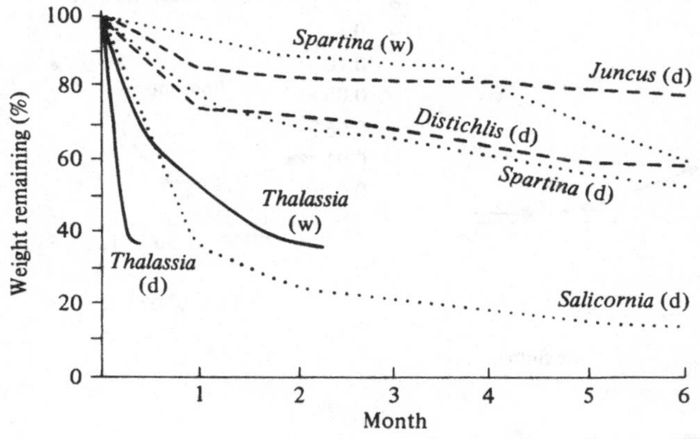

reed stems, for instance, may accumulate in lagoons, to some extent providing a reservoir of potential food materials. Decomposition rates are reduced quite markedly under conditions of anoxia and so this build-up of material is particularly characteristic of lagoons with similar environments to that described for Swanpool (p. 23, p. 30).

As in estuaries, it would appear that detritus or vegetable debris serves as a food source mainly via the activities of micro-organisms. Flagellates may increase dramatically in numbers at times of sea-grass decay and it seems likely that they are utilising, among other substances, dissolved organic compounds leached from the leaves in the early stages of the decomposition process. In this way, some of the products of decay return to the planktonic system. Bacteria and fungi are responsible for the later stages of decomposition of macrophyte debris, and bacterial densities are of the order of 10^9–10^{10} bacteria per gram (dry weight) of seagrass detritus. Although the bacteria are a source of food to many of the detritus-feeding macrofauna, they are also consumed by other detrital micro-organisms. Heterotrophic flagellates, for example, which probably are very largely dependent on bacteria, occur in sea-grass detritus in densities of $5 \times 10^7 - 10^8$ per gram (dry weight), and ciliates, many of which take bacteria, achieve densities of 10^4–10^5 per gram (dry weight). Just considering the bacteria, zooflagellates and ciliates, and many meiofaunal organisms will also be present, every gram of sea-grass detritus will provide some 9 mg of micro-organisms to consumers of the detrital aggregate (Fenchel, 1977).

Detritus is formed on the bed of a lagoon and therefore the detrital particles are associated with the surface layers of the sediment. There, too, occur numerous species of living diatoms and of other micro-algae, which will also be ingested by consumers of sediment. Yet further micro-algae are associated with the living macrophytes (see e.g., Kikuchi & Pérès, 1977); indeed the biomass of attached algal epiphytes may exceed that of the sea-grass leaves to which they are attached, and their productivity may be very much larger than that of their support. These also will provide a rich food source for consumers; young fish have been observed 'carefully picking epiphytes and [the] accompanying epifauna from [sea] grass blades' (Wood *et al.*, 1969).

4.3.2 *Consumers*

Excluding the top carnivores and the consumers of living phytoplankton, most lagoonal animals consume detritus, benthic algae and epiphytes rather indiscriminately, although living plant material may be preferred. The grey mullets are prime examples. They consume mats of algae on the sediment surface and epiphytic species on the submerged macrophytes when these are available; sift the loose sediments to obtain benthic algae and detritus when they are not; and they will also consume the smaller macrofauna including oligochaetes, chironomid larvae, amphipods, small gastropods and even mysids.

Most lagoonal invertebrates are deposit feeders and/or browsers, several of them with an ability to suspension feed as well. Even though phytoplankton may be abundant, specialist suspension feeders are not usually common (although, for example, sponges abound in some lagoons in the Arabian Gulf). This is probably attributable to the large suspended sediment loads, and several of the suspension feeding bivalves which do occur probably rely in part on surface benthic material temporarily put into suspension. Filter-feeding mussels and oysters are particularly associated with the sea-water influenced zone and with the entrance channel itself, where water velocities are relatively high. Like the other invertebrates, the nektonic prawns are catholic in their food requirements, consuming algae, detritus and smaller animal species.

The vertebrate members of the lagoonal fauna contain some detritus feeders (the grey mullets above) and some planktivores (of which the anchovies are examples), but the majority are opportunistic omnivores or carnivores, taking the invertebrate detritus feeders, prawns, other fish, and (incidentally?) benthic algae and detritus. This applies as much to the various bird species frequenting lagoons (ducks, herons, egrets, spoonbills, pelicans, grebes, gulls and cormorants) as to the fish.

A generalised food web is presented in Fig. 4.4, together with the one constructed by Edwards (1978) for the Caimanero/Huizache Lagoon system.

Fig. 4.4. *a*. Diagrammatic lagoonal food web. *b*. Hypothetical food web for the Caimanero/Huizache Lagoons. After Edwards, 1978.

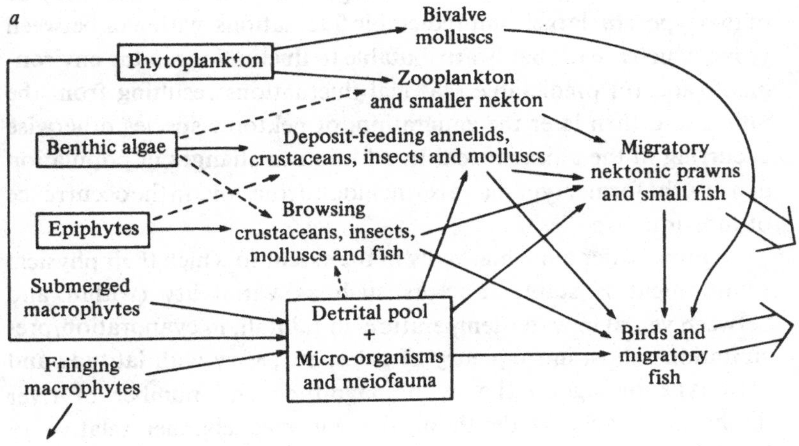

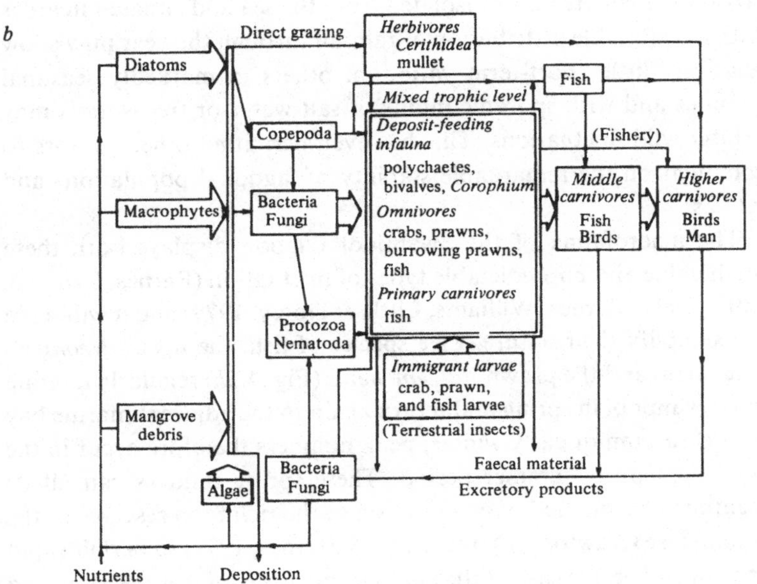

4.4 Seasonal and other population fluctuations

Several authors have commented on the large fluctuations in population density displayed by lagoonal animals. These may be of two types: (*a*) largely unpredictable fluctuations, within or between years, which are probably attributable to fluctuations in the environment; and (*b*) predictable seasonal fluctuations resulting from the influx, and then later the emigration, of nektonic species otherwise occurring in the adjacent sea. In addition to changes in population density, the former type may also include fluctuations in the occurrence of individual species.

Lagoons differ quite markedly in the extent to which their physical environment is stable. Factors such as variability (within and between years) in water temperature, in rainfall, in evaporation/precipitation ratios, in frequency of gales, etc., vary with latitude and with type of lagoon (i.e. with magnitude and number of river discharges, size and depth of the entrance channel relative to lagoonal volume and marine tidal range, etc.). Therefore whilst lagoons which are almost isolated from the sea and situated in areas with an equitable distribution of rainfall through the year may show relatively little short-term variation, others in markedly seasonal climates and with sporadic inputs of salt water or freshwaters may exhibit wild fluctuations. This has even permitted other authors to comment on the remarkable stability of lagoonal populations and faunas!

The macrofauna of the Swanpool Lagoon displays both these predictable and unpredictable types of fluctuation (Barnes, Dorey & Little, 1971; Barnes, Williams, Little & Dorey, 1979) and it will serve to exemplify their nature. Five species of fish, the mysid *Neomysis* (Fig. 4.5*a*) and the prawn *Palaemonetes* (Fig. 4.5*b*) regularly migrate into Swanpool in spring and return again to the adjacent marine bay in late autumn or early winter; peak numbers therefore occur in the June–September summer period. These spring influxes can all be regarded as a method of exploitation of the rich food reserves of the lagoon (see Crawford, Dorey, Little & Barnes, 1979) to permit rapid growth and, in the case of the nektonic crustaceans, for breeding and raising the young. One or two generations are produced in the lagoon per year and growth of *Palaemonetes*, for example, was such as to increase the average weight in 1975 from 0.7 mg in June to 30 mg

Fig. 4.5. Seasonal fluctuations in numbers of nektonic crustaceans in Swanpool, Cornwall. *a*. The mysid *Neomysis vulgaris*. *b*. The prawn *Palaemonetes varians*. Numbers are average catches per standard seine tow. From Barnes *et al.*, 1979.

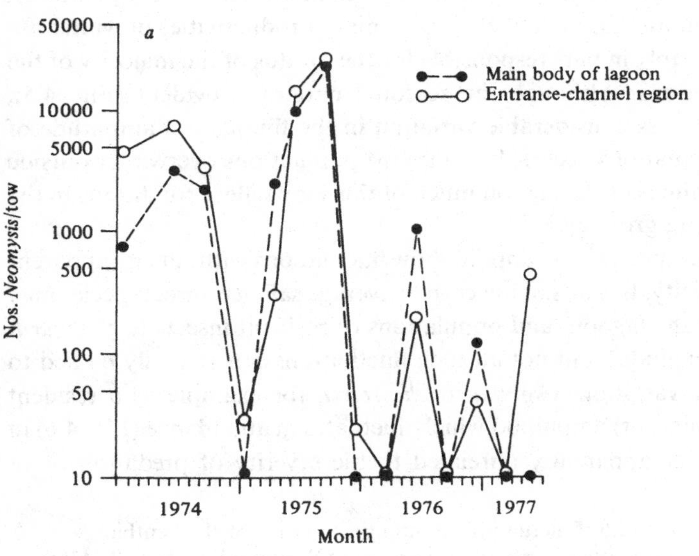

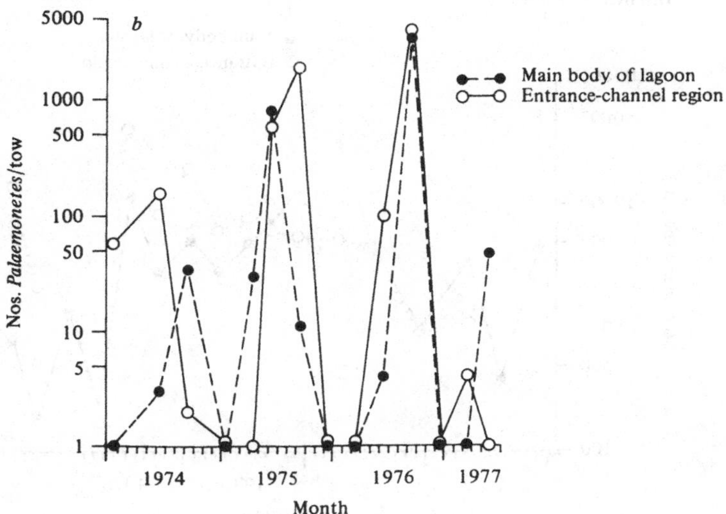

in September, and in 1976 from 1 mg to 10 mg over the same period (dry weights). Why these species should leave the lagoon on the approach of winter is less certain; indeed, a few mysids and prawns do often attempt to overwinter (and succeed) although they only represent a small proportion of the population. The low temperatures, low salinities and relatively low primary productivities of winter are all probably in part responsible for the exodus of the majority of the populations. Although the seasonal pattern is evident (Fig. 4.5), nevertheless considerable variation in the timing and amplitude of the fluctuations occurs, but since the populations overwinter outside the confines of the lagoon much of this may reflect conditions in the wintering grounds.

Other species in Swanpool show fluctuations without any apparent periodicity. In years of lower than average salinity, insect species may invade the lagoon, and populations of resident insects (e.g. *Sigara*) may 'explode', but not all such fluctuations can so easily be tied to salinity variation. *Gammarus chevreuxi*, for example, is a resident (non-migrant) amphipod which fluctuates in abundance (Fig. 4.6) in a manner apparently unrelated to the severity of predation or of

Fig. 4.6. Fluctuations in numbers per m² of the benthic amphipod *Gammarus chevreuxi* in Swanpool, Cornwall. After Barnes *et al.*, 1979.

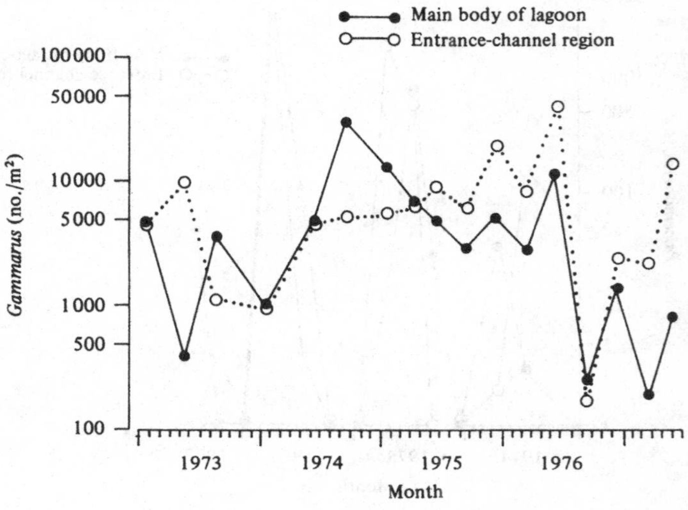

competition, to the numbers being washed out of the lagoon, or to the action of any single environmental variable (Barnes *et al.*, 1979). These authors could only account for the pattern by suggesting that although most environmental conditions were favourable, combinations of high salinity and high temperature adversely affected survival.

The physical environment of Swanpool is one subject to wide variation in temperature, salinity, oxygen, pH, etc., both within and between years, and so it might be supposed that such environmentally-induced fluctuations would there be particularly marked. This may be so, but such instability is a feature of very many lagoons and the populations in each lagoon can be expected to fluctuate in relation to the specific pattern of variation in their individual habitats.

4.5 Biomass, productivity and diversity

Whittaker (1975) provides 'average' values, as dry wt/m^2/yr, for the primary productivity of different ecosystems; these include open sea, 125 g; lakes, etc., 250 g; coastal seas, 360 g, and algal beds, reefs, estuaries, swamps and marshes, 1800–2000 g, the latter category including the highest known productivities of any habitat in the world. Using the same units, the approximate ranges of phytoplanktonic production per year in aquatic systems vary from 15–70 g (oligotrophic lakes and oceans) to 150–500 g (eutrophic lakes and seas). It is against this background that the productivity of lagoons must be judged.

As yet, few studies of the productivity of lagoonal plants have been attempted and even fewer have investigated the productivity of all the photosynthetic organisms or communities in any single lagoon. Representative data are presented in Table 4.1. Many characteristic types of lagoonal vegetation also occur in other habitats, however, and there is no reason to suppose that their productivity in lagoons differs markedly from elsewhere. Thus the productivities of saltmarshes, mangrove-swamps and reedbeds, which elsewhere lie in the range 1000–4000 g dry wt/m^2/yr, will place lagoons amongst the most productive of ecosystems, the more so because of their long shorelines in relation to surface area or volume. Estimates from these other habitats also indicate that 350–> 2000 g dry wt/m^2/yr of this production will be contributed in the form of particulate detritus to the

Table 4.1. *Representative primary productivities of various lagoonal plant assemblages*

Area	Plant type	Productivity[a]	Source
Venetian High Lagoon, Italy	Phytoplankton	294 $g/m^2/yr$ (net) (2.8 $g/m^2/day$ max.)	Vatova, 1960
Selsø Lake, Denmark	Phytoplankton	560 $mg/m^3/hr$ (gross) (max.)	Wøldike, 1973
Swanpool, England	Phytoplankton	10–14 $g/m^3/day$ (net) (max.)	Crawford *et al.*, 1979
Oyster Pond, Massachusetts, USA	Phytoplankton	7.6 $g/m^2/day$ (gross) (max.)	Walsh, 1965
Cochin Backwater, India	Phytoplankton	372 $g/m^2/yr$ (net)	Qasim, 1970
Caimanero/Huizache Lagoons, Mexico	Phytoplankton	4.8 $g/m^2/12$ hr (net) (ave)	Edwards, 1978
Caimanero/Huizache Lagoons, Mexico	Benthic diatoms	230 $mg/m^2/hr$ (gross) (max.)	Edwards, 1978
Caimanero/Huizache Lagoons, Mexico	*Ruppia*	10.1 $g/m^2/hr$ (gross) (max.)	Edwards, 1978
Izembek Lagoon, Alaska, USA	Phytoplankton	160 $mg/m^2/day$ (net) (ave)	Barsdate *et al.*, 1974
Izembek Lagoon, Alaska, USA	*Zostera*	22.4 $g/m^2/day$ (net) (max.)	Barsdate *et al.*, 1974
Swartvlei Lagoon, South Africa	*Chara*	263 $g/m^2/yr$ (net)	Howard-Williams, 1978
Swartvlei Lagoon, South Africa	*Potamogeton*	16 $g/m^2/day$ (net) (max.) (2506 $g/m^2/yr$)	Howard-Williams, 1978
Florida lagoons, USA	*Thalassia*	1200 $g/m^2/8$ mnth (net) (ave)	Wood *et al.*, 1969
Texas and Florida lagoons, USA	Seagrasses	2800 $g/m^2/yr$ (net) (ave)	McRoy & McMillan, 1977
Florida lagoons, USA	Epiphytes	400 $g/m^2/yr$ (net) (ave)	McRoy & McMillan, 1977
Bahia Concepcion, Mexico	Phytoplankton	8 $mg/m^3/h$ (net) (ave)	Gilmartin & Revelante, 1978
Estero de la Lobos, Mexico	Phytoplankton	106 $mg/m^3/h$ (net) (ave)	Gilmartin & Revelante, 1978
Laguna de la Cruz, Mexico	Phytoplankton	162 $mg/m^3/h$ (net) (ave)	Gilmartin & Revelante, 1978
Estero de Urias, Mexico	Phytoplankton	272 $mg/m^3/h$ (net) (ave)	Gilmartin & Revelante, 1978

[a] Original data in g (carbon) have been converted to g (dry wt) using a factor of 2 for algal cells and one of 2.8 for macrophytes; all units are in g or mg dry wt.

Table 4.2. *Representative lagoonal biomass values*

Area	Organism	Biomass	Source
Selsø Lake, Denmark	Phytoplankton	130 mg Chl. *a*/m^3 (max.)	Wøldike, 1973
Swanpool, England	Phytoplankton	540 mg Chl. *a*/m^3 (max.)	Crawford *et al.*, 1979
Étang de Bages-Sigean, France	Phytoplankton	> 12 mg Chl. *a*/m^3 (max.)	Jacques *et al.*, 1975
—	Small sea-grasses	100–200 g dry wt/m^2 (ave) (500 g dry wt/m^2 max.)	McRoy & McMillan, 1977
—	Large sea-grasses	0.5–1.5 kg dry wt/m^2 (ave) (8 kg dry wt/m^2 max.)	McRoy & McMillan, 1977
Florida, USA	Epiphytes	370 g dry wt/m^2	McRoy & McMillan, 1977
Caimanero/Huizache Lagoons, Mexico	*Ruppia*	1 kg dry wt/m^2 (max.)	Edwards, 1978
Caimanero/Huizache Lagoons, Mexico	Gastropods	30 g dry wt/m^2	Edwards, 1978
Caimanero/Huizache Lagoons, Mexico	Polychaetes	8 g dry wt/m^2	Edwards, 1978
Caimanero/Huizache Lagoons, Mexico	Amphipods	6.5 g dry wt/m^2	Edwards, 1978
Swanpool, England	*Gammarus*	7.3 g dry wt/m^2 (max.)	Barnes *et al.*, 1979
Upper Laguna Madre, Texas, USA	Fish	22–392 kg wet wt/ha	Hellier, 1962
Northern Sivash, Sea of Azov, USSR	Invertebrates	0.2–0.3 kg wet wt/m^2 (ave)	Zenkevitch, 1963
Dybsø Fjord, Denmark	Invertebrates	75–112 g wet wt/m^2	Muus, 1967
Dievengat, Belgium	*Nereis*	38.5 g dry wt/m^2 (max.)	Heip & Herman, 1979

lagoon as a whole. In the Mississippi Delta, Louisiana, a region directly comparable with many lagoonal areas, Hopkinson, Gosselink & Parrondo (1978) and White, Weiss, Trapani & Thien (1978) have recently recorded net productivities for a variety of fringing marsh plants in the range 1160–6040 g dry wt/m^2/yr, with loss rates of 300–5140 g dry wt/m^2/yr. For the reed, this represents an export of up to 4.9 mg/g/day. Mean above-ground biomass was within the range 390–900 g dry wt/m^2, with maxima up to 2190 g/m^2.

In addition, productivities of benthic diatoms, etc. in comparable habitats are 350–1150 g dry wt/m^2/yr, and of submerged sea-grasses are 1000–4000 (up to a maximum of 8000) together with a further contribution of some 400 g dry wt/m^2/yr from their epiphytes. Clearly, then, all the evidence indicates that lagoons are characterised by an exceptionally high productivity.

The relationship between productivity and biomass in ecosystems is variable and depends upon the growth and multiplication rates of the more important organisms: small organisms tend to have small standing stocks but large productivities, and large organisms the converse. Most lagoonal organisms (excluding the fringing vegetation) are relatively small, but nevertheless lagoons often also support fairly large total biomasses, with, for example, (dry weight) quantities of submerged macrophytes in the order of 300–3000 g/m^2 (Table 4.2). Not surprisingly, perhaps, in view of the abundance of potential food, the polychaete *Nereis diversicolor* is a species which achieves a larger individual density, biomass and productivity in lagoons than in estuaries.

These high biomasses and high productivities are attained, however, by a fauna and flora of relatively low diversity. Few species are lagoonal. Again, little work has been devoted to comparative lagoonal diversity, although Peterson (1975) has investigated this, and community constancy, in two Californian lagoons, the Mugu Lagoon and the Tijuana Slough. The environments studied in the two lagoons were very similar: both were areas of sandy bed in the marine-influenced zone (salinity 34‰), with comparable tidal ranges, water velocities and temperature regimes. Occasional heavy rainstorms, however, were more frequent and more severe in Mugu Lagoon.

Only the larger benthic macrofauna (> 3.2 mm) was sampled and this comprised many bivalves, some gastropods, crustaceans and echinoderms, and a brachiopod. Individual animals were some four times more abundant in Mugu, and 46 species were represented there, in comparison to 32 species in Tijuana Slough (27 were common to the two lagoons). In spite of the richer fauna, the species diversity of the Mugu fauna was significantly lower than that in Tijuana Slough (a consequence of the more equitable relative densities). Further, the species diversity and total number of species obtained per series of samples varied less at Tijuana than in Mugu Lagoon; the measure used to indicate degree of temporal variation was less than one-fifth of that for Mugu.

The lower species diversity and greater temporal variability of the Mugu fauna was attributed by Peterson to the effects of the more severe rainstorms in occasionally, but drastically, reducing the salinity of the lagoonal water. It was also considered possible that the greater spatial heterogeneity of the Tijuana habitat contributed to the higher diversity and smaller population fluctuations there. The factors influencing the higher density of individuals and the richer fauna in Mugu Lagoon were not specifically investigated; although, since many of the Mugu species were rare, the difference in total number of species may have been more apparent than real.

Several lagoons occur in groups, with different lagoons containing very different faunas and floras (e.g. those of the Gulf of California, Mexico, investigated by Gilmartin & Revelante, 1978); these provide an ideal opportunity to employ a (much needed) comparative approach to the nature of the factors influencing lagoonal ecosystems. Investigation of this field has hardly begun, but already studies of lagoonal diversities have contributed to diversity theory in general (see Young & Young, 1977).

5

Strategies for lagoonal species

5.1 Introduction

Hedgpeth (1957), Muus (1967) and Schachter (1969) have drawn attention to the existence of species pairs (or trios) in the related coastal marine, estuarine and lagoonal habitats. Thus one may have a group of closely-related ('sibling') species which replace each other in the different systems (Table 5.1). The question therefore arises why does such and such a species inhabit lagoons whereas its sibling occurs in estuaries or in the adjacent sea? To answer this, one requires comparative data on the life-history (and other) strategies of all the species in the complex, and regrettably this is very rarely available. Indeed, at the moment, it is probably impossible to answer the question with any degree of certainty; however, by examining in some detail two species-complexes from the relatively well-known

Table 5.1. *Some sibling-species groups in the coastal habitats of north-west Europe*

Coastal marine	Estuarine	Lagoonal
Cerastoderma edule		*C. glaucum*
Hydrobia ulvae		*H. ventrosa* and *H. neglecta*
Littorina saxatilis rudis[a]		*L. s. tenebrosa*[a]
Idotea balthica		*I. chelipes*
Sphaeroma serratum	*S. rugicauda*	*S. hookeri*
Nereis virens	*N. diversicolor*	
Pomatoschistus minutus	*P. microps*	

[a] The systematics of *Littorina saxatilis* is currently in considerable confusion; some authors regard it as a single, highly variable species, others recognise three, four or more species within the aggregate. The usage here follows that of, e.g., Muus, 1967.

northern European coastal fauna, it may be possible at least to uncover some clues. The two complexes are: *Hydrobia ulvae* versus *H. ventrosa* and *H. neglecta*; and *Cerastoderma* (or *Cardium*) *edule* versus *C. glaucum* (= *C. lamarcki*) and in each case, the first named is coastal marine and estuarine, whilst the second is typically lagoonal.

Clearly, one will expect each species generally to be adapted to its preferred habitat; for example *C. edule* appears able to withstand relatively heavy wave action and is adapted to tidal conditions, whereas *C. glaucum* is adapted to quieter non-tidal circumstances. But which is cause and which is effect? Morphological adaptations of the shell, for example, may be merely niceties or fine adjustments evolved after the two species had 'settled' in their respective habitats. Indeed, populations of *C. glaucum* displaced into estuarine habitats develop shells more akin to that of *C. edule*, as seen in the Southampton Water populations. It must be from rather more fundamental biological attributes, such as competitive abilities, reproductive strategies, growth and feeding strategies, etc., that the solution to the problem should be gleaned. Such an approach should eventually also tell one what the various physical differences between estuaries and lagoons actually mean in biological terms.

5.2 Example 1: *Cerastoderma glaucum*

Most of our knowledge of the biology of the lagoon cockle, *C. glaucum*, in relation to that of its sibling *C. edule*, stems from the work of Dr C. R. Boyden (see Boyden, 1972, 1973; Boyden & Russell, 1972 and earlier work cited therein). *C. edule* and *C. glaucum* are evidently very closely related; indeed they can be hybridised in the laboratory and the hybrids produced are fertile. Appreciation of their distinctiveness only really dates from 1958 and earlier records of the two are usually under the one name *C. edule*. Thus the copious records of '*C. edule*' from the Baltic, Mediterranean, Black and Azov Seas and their subsidiary lagoons probably all refer to *C. glaucum*. *C. edule* itself is found in fully marine and in estuarine habitats, although in the latter it may, rarely, occur with the estuarine form of *C. glaucum*. Boyden & Russell (1972), for example, surveying the British and Irish distribution of *C. glaucum* found it in 43 lagoons but in only six tidal areas, in each of which *C. edule* was also present.

These mixed populations constitute a particular problem in that individuals intermediate in morphology occur in them; as for example in the Isefjord/Roskilde Fjord lagoon complex in Denmark (Rasmussen, 1973). These apparent intermediates may be actual hybrids or they may be individuals of both species which have convergently evolved a very similar shell form in response to the same habitat. In fine sediments, for instance, *C. edule* possesses a very

Fig. 5.1. The morphology of the cockles *Cerastoderma glaucum* (*a*) and *C. edule* (*b*). The ratio *y*:*x* is one of the more reliable means of identification; in *C. glaucum* it is 1:4–5, in *C. edule* 1:3. From Muus, 1967.

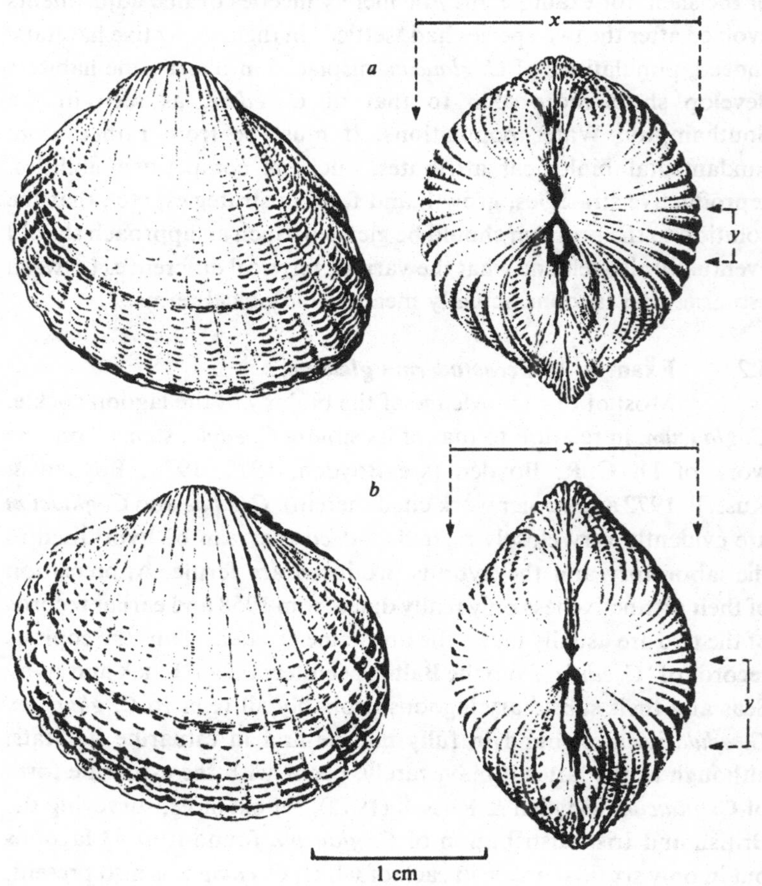

Table 5.2. *The preferred habitats in Britain of* Cerastoderma edule *and* C. glaucum. *After Boyden & Russell, 1972*[a]

	C. edule	*C. glaucum*
Habitat	Estuary	Lagoon
Temperature (°C)	3–20	0–25
Salinity (‰)	15–35	5–40
pH	7.5–8.5	7.7–10.1
Oxygen (% saturation)	90–105	0–200
Substratum	Mud, sand	Mud, sand, shingle, vegetation
Exposure	Sheltered	Extremely sheltered
Tidal range (m)	1.5–10	0–3.0
Exposure to air (% time)	0–50	0–5
Habit	Buried in sediment	Buried, on surface, or suspended in vegetation

[a] Boyden & Russell also provide a useful list of many lagoonal habitats in Britain and Ireland.

glaucum-like shell. The differences between the typical forms of the two species are shown in Fig. 5.1.

Not surprisingly, considering the close relatedness of the two, *C. glaucum* and *C. edule* share many biological attributes. Their reproductive patterns, for example, are very similar, although *C. glaucum* appears to require a higher temperature for spawning (*c.* 20 °C in British populations *c.f.*, 13 °C for *C. edule*), and their larvae are smaller and have a shorter planktonic life (only 1 week as opposed to 2–6 weeks). The larvae will settle on a variety of substrata, but one of the more obvious ecological features separating the cockles is the preference of larvae of *C. glaucum* for vegetation (*Zostera*, *Ruppia*, *Chara*, etc.). Young *C. glaucum* characteristically climb on submerged macrophytes using their byssus threads. When older they may descend to the substratum, but not infrequently large lagoon cockles may still be found in masses of green algae such as *Chaetomorpha*. Even when free of the submerged vegetation, *C. glaucum* may live on the surface of, rather than in, the sediments. The common cockle, *C. edule*, never climbs up macrophytes (see Table 5.2) although, rarely, it may be found in filamentous algae.

Although the two cockles differ markedly in the salinity range over which they occur naturally, *C. glaucum* inhabiting the wider range (4‰–> 100‰), the major habitat difference is the dependence of *C. edule* on a vigorous tidal regime and of *C. glaucum* on highly sheltered conditions. Even within a lagoon, maximum densities of *C. glaucum* are found in the most sheltered localities, and in these fairly stagnant areas intra-specific competition for food (mainly temporarily suspended benthic diatoms) is probably intense. Owing to their ecological separation, it is difficult to assess the magnitude of interspecific competition between the two. In the mixed population in the Crouch Estuary, Essex, Boyden (1972) showed that their life-spans were similar, as were their growth rates, although since *C. edule* spawned 7–8 weeks earlier in the year they had a 'head start' and, age-for-age, *C. glaucum* were smaller. Perhaps in most mixed populations in estuarine habitats, *C. glaucum* is in the minority and recruitment is often more sporadic, and this may be as a result of interspecific competition. It may, however, equally be a result of the recent displacement of *C. glaucum* into comparatively unfavourable habitats (from the point of view of its feeding ecology) by land reclamation. Boyden suggests that many of the localities in south-eastern England in which *C. glaucum* occurs are but remnants of a once more widespread lagoonal habitat fragmented by natural and artificial reclamation. Reproductive isolation was effected in the mixed Crouch population by different spawning periods: *C. edule* in May and *C. glaucum* in July. In lagoons, *C. glaucum* may spawn earlier in the year; but this may merely reflect the fact that shallow, semi-stagnant lagoons attain higher temperatures earlier in the year than do well-flushed estuaries.

Recently, Brock (1979) has suggested the means by which the movement of water may affect the cockles. In estuarine/lagoonal environments in Denmark, Brock found that *C. edule* grew faster than *C. glaucum* (*c.f.* above) and that it could not survive periodic shortages of food. It was therefore restricted to habitats in which relatively strong tidal movements continually supplied food, even if this was only surface deposits whirled into suspension. In contrast, the slower growth rate of *C. glaucum* imposed no such requirement and it could inhabit more stagnant areas. Even so, its climbing behaviour can be interpreted as serving to bring it nearer to its planktonic food source, feeding conditions for suspension feeders being extremely poor at the base of dense stands of vegetation because of the negligible flow rates of the water. The limited burrowing abilities of *C. glaucum* also ill-adapt it for life in the loose sands typical of areas with tidal water fluxes.

Considering the two species, one is left with more of an impression of their similarity than of differences, as perhaps befits a recently established species-pair. The wider tolerance of many environmental variables, the preference for submerged macrophytes and that for shelter appear the only lagoonal adaptations of *C. glaucum*, although some features of their larval biology may also be relevant (see pp. 62–64). The two small cockles *Cardium exiguum* (marine) and *C. hauniense* (lagoonal) are in several respects a comparable species-pair.

5.3 Example 2: *Hydrobia ventrosa* and *H. neglecta*

Four small hydrobiid snails are abundant in north-west European estuaries and lagoons: *Hydrobia ulvae*, *H. ventrosa*, *H. neglecta* and *Potamopyrgus* (often recorded as *Hydrobia*) *jenkinsi*. These have been studied intensively in Denmark by Muus (e.g. 1967) and by Fenchel and his co-workers (Fenchel, 1975*a*,*b*; Fenchel & Kofoed, 1976; Hylleberg, 1975; and references cited therein). The snails show many similarities to the picture obtained in respect of the

Fig. 5.2. The morphology of estuarine and lagoonal species of *Hydrobia*: *H. ventrosa* (*a*); *H. neglecta* (*b*); *H. ulvae* (*c*). From Muus, 1967.

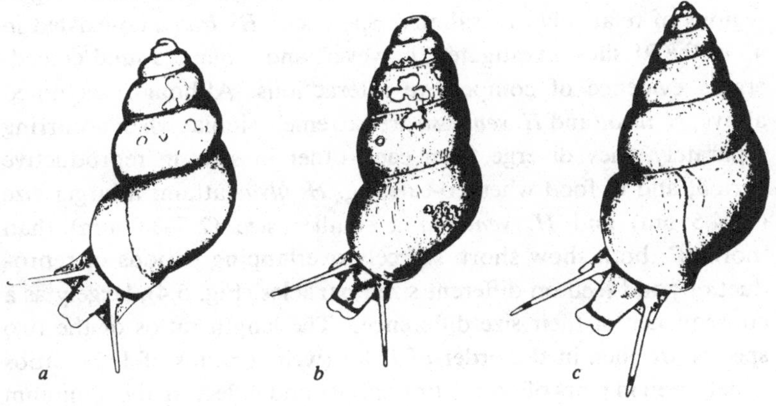

cockles: they are morphologically extremely similar (Fig. 5.2); they are differentially distributed along the gradients of salinity and water movement (Fig. 5.3); some species show a preference for submerged vegetation; and two or more species may sometimes be found in the same habitat.

Biologically, there are particularly few differences between the three *Hydrobia* species. When occurring separately, they are of similar size, attain the same age, and feed on the same types of food particles (diatoms, bacteria, etc.). Such differences as there are relate to their salinity optima, their competitive abilities and their reproductive strategies. *H. ulvae* has a salinity optimum of 30‰ and feeds maximally at salinities of 20–30‰ dependent on temperature. *H. ventrosa* is the most catholic in its salinity requirements, but tends to show an optimum in lower salinities than *H. ulvae* (*c.* 20‰). Its feeding activity is strongly influenced by temperature, being maximal at various points in the range 10–20‰ under environmental conditions of 5–35 °C. The preferred range of *H. neglecta* is sandwiched between those of its two congeners, with an optimum at 25‰, although this decreases a little with increase in temperature. The order of increasing salinity optima in these three species is paralleled by those of increasing tolerance of H_2S, anoxia and desiccation.

Fenchel (1975*a,b*) studied the distribution of all four hydrobiids in the large estuarine/lagoonal complex of the Limfjord, Denmark. In general, their distribution conformed to that presented in Fig. 5.3*a*, with *H. ulvae* dominating open areas, *H. ventrosa* and *H. neglecta* being characteristic of lagoonal zones, and *P. jenkinsi* occurring in regions of relatively low salinity. Species of *Hydrobia* co-existed in 43 of the 91 sites investigated, however, and Fenchel found considerable evidence of competitive interactions. Although, as noted above, *H. ulvae* and *H. ventrosa* are extremely similar when occurring separately, they diverge from each other in size, in reproductive period, and in food when co-existing. *H. ulvae* attains a larger size (*c.* 4.5 mm) and *H. ventrosa* a smaller size (2.5–3.0 mm) than 'normal'; both show short, scarcely-overlapping periods of reproduction; and feed on different sized particles (Fig. 5.4), largely as a consequence of their size differences. The length ratios of the two species are then in the order of 1:1.3 (weight ratios of 1:2), ratios widely seen in pairs of co-existing species and reflecting the minimum

Fig. 5.3. The distribution of Danish hydrobiids in relation to salinity and to exposure/shelter. From Muus, 1967. *a.* Diagram of optimal ranges. *b.* Distribution along the salinity gradients of Randers Fjord (upper frame) and Kysing Fjord (lower frame). (Broken lines, sporadic; complete lines, common; thickened lines, numerous, i.e. > 2000/m².)

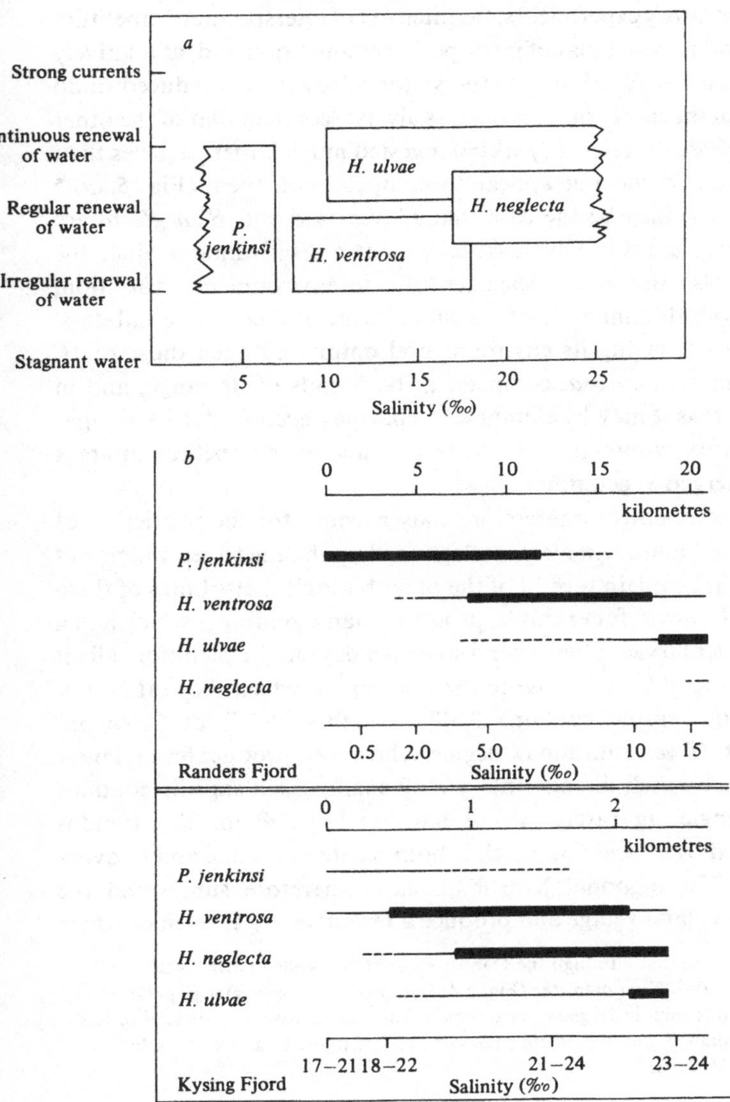

size difference necessary to permit utilisation of different resources, thereby preventing competitive exclusion. Other factors do also influence size (amongst them being sediment type), but Fenchel considered that the presence or absence of interspecific competition was overwhelmingly the most important variable.

In laboratory experiments, the intensity of interspecific competition was found to equal that of intraspecific competition and, at relatively high salinities, *H. ulvae* was the victor. Growth was reduced in all cases, but mortality of *H. ulvae* was always less than that of the other two *Hydrobia* species. *P. jenkinsi* ingested much larger particles than *Hydrobia* and did not appear to compete with them (Fig. 5.4*a*).* Therefore it may be the case that *H. ventrosa* and *H. neglecta* are confined to areas in which *H. ulvae* is at a disadvantage. Since the latter is also the species least sensitive to environmental variation, it therefore dominates both coastal marine and estuarine habitats. *H. neglecta*, having its environmental optima between those of *H. ulvae* and *H. ventrosa*, competes at both ends of its range, and in several areas it may be eliminated. This may account for its comparative rarity, although it is possibly a superior competitor in areas of submerged vegetation.

Such competitive interactions may account for the restriction of *H. ventrosa* and *H. neglecta* to lagoons hypohaline to sea water, but they do not explain several of the other biological attributes of these species. *H. ulvae*, for example, produces many small eggs which hatch into veliger larvae. These spend some ten days in the plankton, albeit that they tend to stay close to the bottom, on which they often rest. Each egg capsule contains 3–40 eggs (ave 7–25) of 70–90 μm diameter. *H. ventrosa* and *H. neglecta*, however, produce fewer, larger eggs which hatch directly into young snails. Each capsule contains only a single egg (rarely two) of diameter 150–250 μm. This trend is continued by *P. jenkinsi* which is both parthenogenetic and ovoviviparous. The lagoonal hydrobiids have therefore suppressed the planktonic larval stage and produce a few offspring to each of which

* Note that although the Danish *P. jenkinsi* ingested particles of 100–200 μm diameter (Fig. 5.4*a*), studies on a freshwater population of this snail in England have shown that there it feeds on particles of less than 70 μm in diameter, the same size of particles as are taken by *Hydrobia*.

Fig. 5.4. Size of food particles taken by Danish hydrobiids. *a*. Size of food particles in relation to size of snail. Note that *P. jenkinsi* takes larger food particles than *Hydrobia*, and that the three *Hydrobia* spp., size for size, consume the same sized particles. *b*. Size of food particles consumed by *H. ulvae* (open circles) and *H. ventrosa* (filled circles) when occurring together (1) and when occurring separately (2, 3). From Fenchel, 1975*b*.

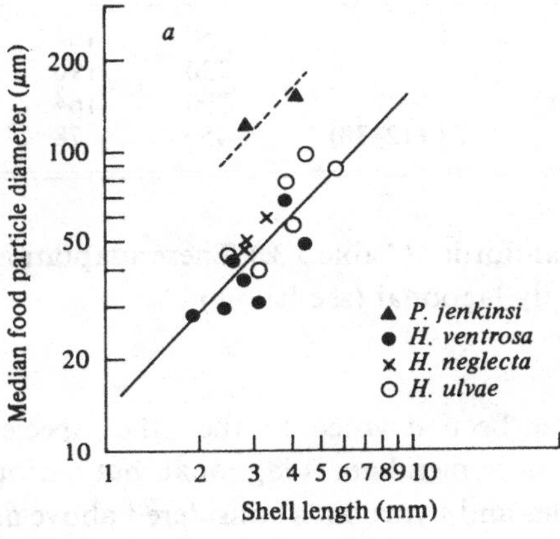

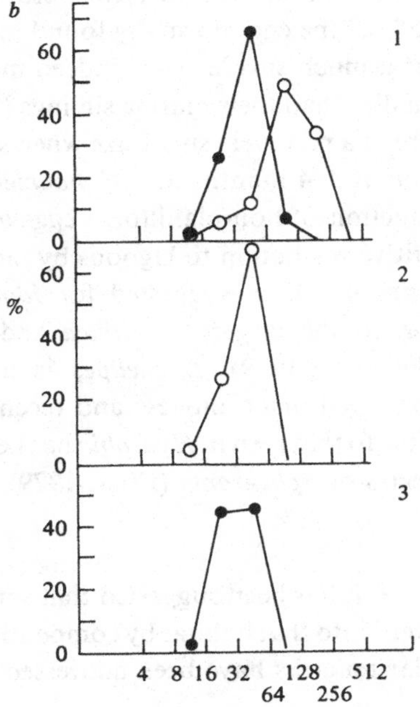

Table 5.3. *Reproductive output of hydrobiid snails from various Danish localities. After Lassen, 1979 and Lassen & Clark, 1979*

Species	Nos. egg capsules (mean and range) per year	Nos. eggs per capsule (mean and range)	Total nos. eggs produced (mean)	Ave egg diameter (μm)
H. ventrosa	50 (24–105)	1	50	164
P. jenkinsi	—	—	230	190
H. neglecta	290 (104–467)	1	290	164
H. ulvae	150 (81–186)	23 (12–38)	3450	78

greater parental investment is afforded (Table 5.3.). These adaptations can be regarded as specifically lagoonal (see below).

5.4 Other examples

Much less work has been devoted to the other species-complexes of which one or more members is lagoonal, but various of the attributes of the cockles and hydrobiids considered above are also seen in other lagoonal animals. *Littorina saxatilis tenebrosa*, for example, produces only some 15% of the eggs normally found in *L. saxatilis rudis*, and it matures at a much smaller size. Indeed most lagoonal species appear to be smaller than their marine siblings (the lagoonal cockles may also mature at a relatively small size when still comparatively young, *C. glaucum* at 3–4 months and *C. hauniense* at only 2 mm length). The production of non-planktonic eggs and larvae is also the norm. Competitive restriction to lagoons by more vigorous marine/estuarine siblings has been suggested for *Idotea chelipes* (=*I viridis*), in relation to the larger *I. balthica* and *I. granulosa*, for example (see Salemaa, 1979). *I. chelipes* is also particularly associated with submerged macrophytes; and recently character displacement comparable to that seen in *Hydrobia* has been demonstrated in the various species of *Sphaeroma* (Frier, 1979).

5.5 Conclusions

It comes as no surprise that it has been suggested that some lagoonal animals have been 'driven' into that habitat by competition from more marine species: similar remarks have been addressed to the estuarine fauna. But the loss or reduction of the planktonic larval phase is, at first sight, more puzzling. After all, it is surely in estuaries, with their comparatively fast flushing rates, that planktonic larvae would be disadvantageous. Larvae of specialist estuarine species could be swept out of the estuary into the sea. It is in these terms that the bentho-planktonic larvae of *H. ulvae* and *Nereis diversicolor* can be explained. Lagoons, with much slower flushing rates and high phytoplankton productivity, would seem to be ideal habitats for planktonic larval stages. This view, however, neglects two important considerations.

The very isolation of lagoons and (frequently) their small size render dispersal within a lagoon relatively easy, and hence this advantage of planktonic larvae would not accrue. Dispersal between lagoons is much more difficult, and their semi-enclosed state and the lack of tidal inputs may mean that prodigious numbers of larvae would have to be produced in order that some should be washed into a hitherto-uncolonised lagoon, and alternative means of dispersal would be more advantageous. These may include transportation in filamentous algae attached to birds' feet.

Ecologists differentiate between two extreme strategies. One is the production of very many small propagules, each of which stands a minute chance of successfully establishing itself, but by virtue of their sheer abundance and wide dispersability, the chance of one or more finding a suitable habitat is large (the *r*-strategy). The other strategy involves the production of few offspring, to each of which can be granted a measure of parental care or investment increasing the chance of their individual survival (the *K*-strategy). The former is associated with species inhabiting temporary environments which, for a time, are competition-free, and the latter with habitats in which competitive interactions are severe. Lagoons are more stable environmentally than estuaries in a number of respects (pp. 20–32), and it has also been suggested that, because of their non-tidal state and the consequent unimportance of wading birds, predation pressure is less intense than in estuaries. Populations of lagoonal invertebrates can therefore more often attain the carrying capacity of their habitat, and for longer periods, and thus competition for food resources (etc.) will be intense. On this basis, lagoonal species should produce offspring which can hold their own, competitively, from the moment of their release; larvae are thus born at an advanced stage, when relatively large, and often with food reserves. They are *K*-selected in comparison to the more *r*-selected estuarine species.

K-selected species, however, are usually relatively large, slow-growing organisms, which breed when relatively old and thereafter produce few young every year for a number of years. *Hydrobia ventrosa* certainly grows slowly in comparison to other species of its genus, but most lagoonal invertebrates have short lives (about $1\frac{1}{2}$ years), breed when young and then die, typical *r*-strategies. This is clearly a paradox, which may be resolvable in terms that although lagoons are stable in the short term (periods of months), they are not so over longer periods. *Cerastoderma glaucum*, for example, although occurring for part of the year in densities resulting in intraspecific competition for food, suffers severe mortality in Danish lagoons during the winter when ice may form. In markedly seasonal climates, the influx of freshwater in the wet season may similarly cause heavy mortality.

Lagoonal species therefore need either to be able to migrate out when conditions are adverse (like the nektonic species) or else to compromise between *r*- and *K*-strategies. In favourable periods, the young stages will be born into a highly competitive situation, but populations must also be able to build up rapidly after environmentally-induced crashes. Both are possible with a strategy combining a *K*-selected reproductive system and an *r*-selected life span, i.e. by living short lives (and therefore being small in size) and by breeding as soon as possible.

Recently, Parker & West (1979), studying the lagoonal mysid *Neomysis vulgaris* (=*N. integer*) in the brackish Lough Furnace in Ireland, have arrived at a similar interpretation of the biology of this nektonic species. The reproductive strategies of mysids in general are *K*-selected, and in that respect they are pre-adapted to lagoonal life, but the rigours of the lagoonal environment have necessitated the evolution of *r*-strategies in most other aspects of the biology of *N. vulgaris*. As Parker & West point out, the compromise between the two strategies is far from perfect, however, and so the species is particularly susceptible to large fluctuations in population density (see pp. 46–48). Being nektonic, *Neomysis* can escape (at least partly) from adverse conditions by migration, although it does not appear to be migratory in Lough Furnace.

Much ecological theory is still in a state of flux, and an alternative explanation of the reproductive strategy of lagoonal animals is possible. Van Valen (1971) has shown that the classic *K*-strategy incorporates – or confuses – two separate phenomena. Species living in highly stressful situations (e.g. deserts, caves and, in aquatic habitats, regions with little oxygen or food materials, or with highly variable salinity) show similar strategies to those inhabiting environments in which competition is intense. Both grow slowly, especially when adult, and produce few, well-provisioned young. These two separate strategies, which have been termed the 'stress-tolerant' and the 'biologically competent', differ in their competitive abilities, in their ability to recolonise denuded habitats, and perhaps in their susceptibility to predation. Stress-tolerant species compete relatively ineffectively, recolonise slowly, and possess few anti-predator devices; intense predation and competition are not features of their severe environments. An apparent *K*-strategy may therefore be related not to severity of competition but to individual persistence in the face of either environmental or biotic adversity.

Some features of lagoonal species can be taken to indicate a stress-tolerant strategy rather than one associated with intense competition, but much further research will be required before lagoonal animals can safely be allocated to one or other of these two *K*-strategy life-styles. Indeed, it is likely that the factors responsible for the evolution of the characteristic reproductive strategies will vary with the nature of the lagoon since, as we noted on p. 46, the extent to which lagoonal environments and populations fluctuate varies. In the present state of our knowledge, it seems safe to conclude only that individual lagoonal animals show a mixture of *r*- and *K*-strategies, and that the importance of competitive interactions will be one of many variables affecting their lives. Our knowledge of estuarine species is only slightly more complete.

The present state of the art is also such that any one set of data is open to differential interpretation. Lassen (1979), for example, examined the relative effort devoted to reproduction by the four Danish hydrobiids and concluded that *P. jenkinsi* and *H. neglecta* both apportioned a larger percentage of their available energy to egg production than did *H. ulvae*. In this sense it is they who are the more *r*-selected. This only serves to emphasise that the various overall strategy types which can be categorised are 'rag-bags' and that different aspects of a single biological activity, like egg production, will be subject to different selective pressures. Although, for example, the production of few large offspring may generally be correlated with the expenditure of a relatively small percentage of energy per unit time on reproduction, there is no necessary reason why they should be linked, and perhaps lagoons provide one example of a habitat in which they are not. Further, Zimmerman, Gibson & Harrington (1979) in their study of the Indian River Lagoon in Florida have questioned whether competition for food is ever likely to occur in lagoons, and have suggested that the apparent partitioning of the total food resource between the various consuming species (which would normally be taken to indicate the effects of interspecific competition) may in fact be an illusion.

This debate on the importance of competition in lagoons exactly parallels that on its importance in estuaries. One view (see, e.g., Beukema, 1976) stresses that estuaries are areas of resource superabundance (their sediments are always rich in organic matter) and that competition is therefore unlikely to occur, whilst the other (see, e.g., Levinton, 1972) documents a wide range of evidence indicating that competition is intense and that food is limiting. It is unlikely that this debate will end until we know much more about what the various detritus feeders of lagoons and estuaries actually assimilate from their food intake in nature.

6

Human use of lagoons

Man's reactions to natural ecosystems are usually ambivalent, and his attitude to lagoons is no exception. On the one hand, he seeks to utilise their high productivity via fisheries, whilst on the other he uses them as convenient receptacles for his wastes.

6.1 Exploitation of resources

The use of lagoons as naturally-occurring fish ponds has a long pedigree. From prehistoric times through to the nineteenth century, the Hawaiian Islanders, for example, used coastal lagoons for the raising of selected fish species (particularly mullet). One of the four types of fish pond for which there is archaeological and cultural evidence, the *loko pu'uone*, was specific to lagoons enclosed by barriers of coral debris and sand. Exchange of water may originally have been via percolation through the permeable barrier, but frequently ditches one or two metres wide were dug to enable more direct access by sea water and its contained organisms. Specially constructed wooden grilles were placed across these ditches to

Fig. 6.1. *Loko pu'uone* in the Hawaiian Islands. *a*. Distribution. *b*. Nature. After Kikuchi, 1976.

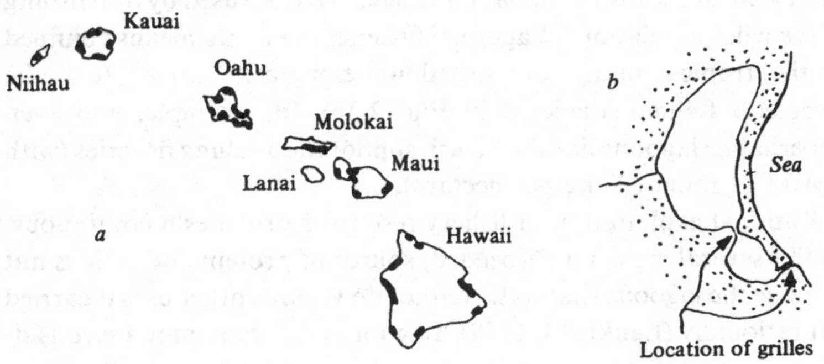

prevent the escape of large fish (Fig. 6.1), whilst permitting young fish to enter; and the lagoons were fertilised by the addition of cut grass and organic marine debris. Yields, in prehistoric times, may have been in the order of 400 kg per hectare (Kikuchi, 1976).

Comparable, or more intensive, exploitation of lagoonal prawns and fish continues today in many parts of the world, both through actual lagoonal fisheries and via fisheries based in the adjacent sea but nevertheless dependent on species which spend parts of their lives in lagoons (see, e.g., pp. 38–39). The potential of lagoon-based fisheries is enormous, especially in the tropics. Most fishing methods are still variants of the 'hunting of wild animals' approach, however, rather than of intensive husbandry or culture, and therefore yields are very low in comparison to that theoretically possible. In the Caimanero/Huizache system, for example, a good annual catch of prawns (*Penaeus vannamei*) would be in the order of $12–14 \times 10^7$ individuals (1200–1400 tonnes), but some $1–2 \times 10^9$ postlarvae enter the lagoon so only about 8.5% are harvested by man (Edwards, 1977). The growth rate of the prawns is phenomenal: about 1.5 mm per day. The exploitation of wild stocks of lagoonal prawns and fish in West Africa and the Indian subcontinent similarly involves losses of up to 95% of the potential catch (of which the proportion formed by prawns varies locally between 20% and 80%). Pillay (1967) estimated that some 9300000 ha of lagoonal and estuarine habitats in Indonesia and the Indian subcontinent are suitable for intensive aquaculture, with potential yields of up to 1000 kg per hectare, although only 140000 ha had been so developed and yields from the unimproved areas were only in the range 30–100 kg per hectare. The 900 (dry season) – 1150 (wet season) km^2 Chilka Lagoon in Orissa has a yield at the lower end of this range, in part caused by overfishing of the wild populations. Lagoonal fisheries are by no means confined to the tropics: mullet are fished in temperate lagoons (e.g. the Orbetello Lagoon depicted in Fig. 2.3*b*), for example, and even hyperhaline lagoons like the Sivash support flourishing fisheries (with a yield of some 100 kg per hectare).

Rational exploitation of fishery resources provides a continuous, and in several areas much-needed, source of protein and it does not damage the lagoonal habitat. Neither do various other uses if carried out rationally (Lankford, 1978). Lagoons can accommodate considerable recreational use, for example, in the form of boating, angling, and the enjoyment of nature in all its forms. The smaller examples also have much to teach man about the structure and functioning of ecosystems: their size, relative simplicity of fauna and flora, high productivity, and comparative independence of the adjacent sea are all conducive to scientific study aimed at an understanding of whole systems. If carried to excess, however, all these uses, including scientific and educational study, can damage lagoonal resources beyond the possibility of short-term recovery. An example of recreational use carried to (unnecessary) excess is the complete isolation of small lagoons within concrete perimeters to serve as ornamental ponds in public parks; several examples of these can be seen in resort towns in south-west England.

6.2 Destruction and degradation

Even more damaging are the uses of lagoons as receptacles of noxious substances and as areas of potential land. The range and nature of these activities are very similar to those described earlier for estuaries (Barnes, 1974), but their effects are often more marked because of the slow flushing rates and shallowness of most lagoons. Therefore pollutants reside for longer periods and higher local concentrations are more easily achieved, and land reclamation is easier to accomplish. The many influences of man on a single lagoon have been described in detail by Bellan (1972) for the Étang de Berre near Marseilles.

In or near residential areas, the scenic qualities of small lagoons also encourage encirclement by luxury housing, with, all too often, the installation of sanitation systems discharging directly into the lagoon. The short-term savings associated with this policy may backfire, however, since eutrophication of the one-time scenic attraction then reduces the value of the surrounding property!

Many fewer towns are built around lagoons than adjacent to estuaries, however, and habitat destruction and degradation are mainly associated with the north temperate zones. Most of the world's lagoons are therefore still in a natural state, even if, in some areas, overfished. There are therefore ample opportunities to rectify the existing scientific neglect of these interesting, widespread and productive coastal ecosystems. That this process has already started is evidenced by the formation in the mid 1970s of an Advisory Panel on Coastal Lagoons by UNESCO and SCOR (Scientific Committee on Oceanic Research, International Council of Scientific Unions) although, as will have been evident in this book, very much more remains to be done, especially at the 'grass-roots' level.

7

Methods for the study of coastal lagoons

There are probably more coastal lagoons in the world than there are professional scientists studying them: they are truly a neglected habitat. This chapter aims to provide those without access to complex equipment or to an extensive library of methodology with some rough-and-ready, shoestring-budget means of remedying this neglect. Unfortunately, however, some fields of investigation do require specialist apparatus and, unavoidably, some shoestrings are more expensive than are others.

It is also the case that to obtain answers to many of the questions which one might wish to ask, a period and/or frequency of study beyond the time available for pursuing an investigation may be required. Hence, before embarking on any study, it is necessary to ask such questions as: Will my programme of research in fact tell me what I hope to find out? This may seem obvious when stated baldly, but the question often remains unasked as even reference to the scientific literature will show. Let me illustrate this with an hypothetical example. Two adjacent lagoons differ markedly in their faunas and floras, and one suspects that differential fluctuation of some environmental variable, say salinity, may in part be responsible for generating this pattern. Leaving aside the difficulty of converting correlations into statements of causality, many environmental variables or processes reach critical levels or operate, respectively, only during exceptional weather conditions, often of the type during which most people would wish to confine themselves to their homes. Such freak conditions may also only occur once every two to fifteen years. Shallow lagoons, for example, may be characterised by very low salinities only during heavy storms. It is therefore most unlikely that a sampling programme based on four to ten visits to a lagoon per year over a period of one or two years would ever record these critical periods and, in a sense, taking regular water samples on such a basis

Table 7.1. *Some sources[a] of techniques, apparatus and methodology appropriate to lagoonal studies*

Briggs, D. *Sediments* (0 408 70815 8). Butterworths, 1977
Campbell, R. C. *Statistics for Biologists*, 2nd Edn (0 521 20381 3). Cambridge University Press, 1974
Chapman, S. B. (Ed.) *Methods in Plant Ecology* (0 632 09060 X). Blackwell, 1976
Cox, G. W. *Laboratory Manual of General Ecology*, 2nd Edn (0 697 04665 6). Brown, 1972
Dowdeswell, W. H. *Practical Animal Ecology*. Methuen, 1959
Holme, N. A. & McIntyre, A. D. (Ed.). *Methods for the Study of Marine Benthos* (0 632 06420 X). Blackwell, 1971
Hulings, N. C. & Gray, J. S. *A Manual for the Study of Meiofauna*. Smithsonian Institution, 1971 (Smithson. Contr. Zool. No. 78)
Poole, R. W. *An Introduction to Quantitative Ecology* (0 07 050415 6). McGraw-Hill, 1974
Pugh, J. C. *Surveying for Field Scientists* (0 416 07520 7). Methuen, 1975
Southwood, T. R. E. *Ecological Methods*, 2nd Edn (0 470 26410 1). Chapman & Hall, 1978
Strickland, J. D. H. & Parsons, T. R. *A Practical Handbook of Seawater Analysis. Bull. Fish. Res. Bd Canada*, No. 167, 1968
Wetzel, R. G. & Likens, G. E. *Limnological Analyses* (0 7216 9243 5). Saunders, 1979

[a] These are all general works and none is devoted specifically to lagoons. ISBN codes have been provided wherever possible, but the publisher listed (and hence the code) is that distributing the work concerned in the UK; different publishers may have this right in other areas.

would not provide one with the information with which to confirm or refute one's hypothesis. This does not mean that one should never undertake any study! – merely that one must choose a project that can yield the relevant information in the time available.

A whole – and large – book could be written on suitable methods for lagoonal research and the material which follows cannot be regarded as a blueprint for a complete survey of any given lagoon. Rather, a few simple methods, which the author knows will provide reasonable results, are described and the reader is referred to the sources listed in Table 7.1 for further information on these and many other suitable techniques. Before describing some appropriate methodology, however, two desiderata must be emphasised. First, little will be gained by taking a regular series of samples *for their own sake*. As indicated above, one should first decide what one wishes to find out; plan, in advance, one's strategy; and then draw up a list of the samples or measurements that one must take and with what frequency. Secondly, many lagoons are small and easily disturbed environments, and the reader must also give thought to the effects that the proposed study might have on the habitat. It is by no means difficult to study a small habitat to destruction. It may therefore be necessary to reach a compromise between the number of samples which, for scientific or statistical purposes, one ought to take and that number which the lagoon can withstand without harm to its inhabitants, erring always on the side of conservation.

Having completed a study and analysed one's results (in which process, and indeed during the planning stage, statistics may be very important (see Table 7.1)), it is well worth considering publishing an account in the journal of the appropriate local naturalists' society. It is only by making information on individual lagoons available to the scientific community at large, that a picture of the biology of lagoons in general can be assembled.

7.1 Topographical and background information

The first requirement in many lagoonal studies is to establish the general topography of the area to be investigated. If one is fortunate, recent maps or aerial photographs may be available, but even then the rapidity of change of lagoonal shorelines may render them out of date at least in some details. The outline of a lagoon, the extent of the fringing vegetation, etc. may be established with the aid of a good compass (of the Brunton type) and of a small surveyor's level and graduated staff. Small lagoons can even be charted with a compass, tape measure and a length of string. Reference to old maps and photographs (often housed in local libraries or museums) may then give valuable information on changes in historical time, and reference to local archives may allow one to date the more recent of such changes and to document past patterns of human use of the lagoon.

Depth is most conveniently measured, in all but the shallowest lagoons, from a small inflatable boat mounted with a battery-operated echo sounder, but a leaded and graduated length of string or a graduated pole with a flat plate affixed to its base will also suffice.

Many transects may be required to obtain a true picture of bottom topography.

7.2 Nature of the lagoonal environment

7.2.1 *Sediment type*

Lagoons are floored by soft sediments and therefore a hand-grab (of the type illustrated in Fig. 7.4), if the bottom is gravelly, or a core-tube (a length of plastic tubing, open at each end and of diameter some 6–7 cm) if the sediment is sand or mud, will serve to obtain a sample for analysis. The depth to which the sample should be taken will vary with factors such as the purpose for which it is to be collected (*c.f.* the depth to which animals burrow), the nature of the sediment itself, etc.; but an all-purpose depth would be in the order of 10 cm.

One of the most useful measures of sediment type, one with which many other features often correlate, is particle size. Unless a sediment derives from more than one source (e.g. as in a mixture of wind-blown sand and lagoonal silt), its component particles approximate to a lognormal size distribution; i.e. if the particles are apportioned to various size classes arranged on a logarithmic scale, they conform to a normal distribution with most particles in the middle classes and progressively fewer towards the two extremes of size. The logarithmic scale most frequently used to describe particle size distributions is the phi (ϕ) scale to the base 2, where, if d is the diameter of a particle (in mm),

$$\phi = -\log_2 d.$$

The value and uses of this particular system are described by Briggs (1977) (see Table 7.1.).

The methods of analysis of particle size distribution depend on whether the sediment is relatively coarse (shingle and/or sand) or fine (soft muds). If it is mainly sand or shingle the following procedure can be used.

> A 100–1000 g sample should be dried to constant weight and that value recorded. It is then tipped carefully on to the uppermost of a stack or 'nest' of sieves of decreasing mesh size. Such sieves may be bought or constructed from metal gauze or nylon netting of different mesh sizes, metal for the larger sizes and nylon for the smaller. Wherever possible, the nest of sieves should extend from

(for shingle) 16 mm mesh aperture (-4ϕ), 8 mm (-3ϕ), 4 mm (-2ϕ), 2 mm (-1ϕ), or from (for sand) 2 mm (-1ϕ), 1 mm (0ϕ), 0.5 mm ($+1\phi$), 0.25 mm ($+2\phi$), 0.125 mm ($+3\phi$) to 0.063 mm ($+4\phi$). A catch-pot can be used to collect material passing through the finest mesh.

The sediment sample is dispersed over the coarsest meshed sieve and then the whole nest is agitated mechanically or by hand for some 10 min. Care must be taken not to overload the nest with sediment. The material retained on each sieve and in the catch-pot can then be weighed and recorded as a percentage of the dry weight of the original sample. (If a fine dust escapes during the sieving, the sum of the individual weights will be less than that of the original sample: the difference in weight can then simply be added to that of the material in the catch-pot.)

The silt and clay particles comprising soft muds do not sieve as readily as sands and hence a different method has to be adopted for such sediments: this relies on Stokes's Law relating the settling velocity of a particle in, e.g., water, to its size.

Again, a sediment sample is dried to constant weight and then some 20 g are weighed out and placed in a 500 ml (or 1000 ml) measuring cylinder of the type fitted with a stopper. The cylinder is topped up to the 500 ml (or 1000 ml) mark with distilled water and one or two drops of Teepol or 0.5–1.0 g Calgon are added. The water should be maintained as near as possible to 20 °C (or any constant and known temperature: if one other than 20 °C is adopted, the settling times given below will have to be recalculated from Stokes's Law).

The sediment is now thoroughly suspended in the water column by inverting the cylinder several times, the cylinder is stood upright and a stop-watch started. By withdrawing samples of water from known depths in the cylinder at different time intervals, it is possible to calculate the quantities of sediment in all the ϕ intervals between +5 and +10. Just before the lapse of a time interval as listed in Table 7.2, a 20 ml pipette is lowered very gently into the cylinder so that its mouth is at a depth of 10 cm below the water level, and when the relevant time is reached a 20 ml sample of water and suspended sediment is collected. This can then be discharged carefully into a suitable vessel (e.g. a pre-weighed crystallising dish), dried to constant weight and that weight recorded. This will represent the weight of the stated size fractions (those smaller than the stated ϕ value) in the 20 ml sample, and multiplication by the relevant factor (in this case $500/20 = 25$, or $100/20 = 50$) will yield the quantity of that fraction present in the original sample.

Table 7.2. *The time taken by various particle sizes to settle 10 cm in distilled water at 20 °C*

Particle diameter (mm)	Time		
	Hours	Minutes	Seconds
0.0625 (+4ϕ)			29
0.0313 (+5ϕ)		1	56
0.0156 (+6ϕ)		7	44
0.0078 (+7ϕ)		31	
0.0039 (+8ϕ)	2	3	
0.0020 (+9ϕ)	8	10	
0.0010 (+10ϕ)	32	38	

Thus a sample taken at a depth of 10 cm, 1 min 56 s after the suspension was allowed to stand, would contain all particles of less than 0.031 mm diameter, and one taken at the same depth after a time of 31 min would contain all those less than 0.0078 mm diameter. If only those two samples were taken, the quantity of particles (*a*) greater than 0.031 mm diameter and (*b*) less than 0.031 mm but greater than 0.0078 mm diameter can be calculated by difference. Most analyses will not require the separate estimation of all these different particle sizes: several categories can be lumped together and three pipette samples should be sufficient for most purposes. All individual weights should, as with the sand, be expressed as percentages of the total weight analysed.

Few sediments comprise particles that are all either greater than or less than 0.063 mm diameter, and therefore both methods should be used in conjunction, with the material collecting in the catch-pot (or, preferably, passing the 0.125 mm mesh sieve) being subjected to the setting-time technique.

Thus far we have succeeded in dividing a given sediment sample into a number of size categories forming known percentages of the whole. This may be of limited intrinsic interest, however, and we must now arrange our results in a form suitable for comparative purposes. This involves two procedures.

First, the individual percentages should be listed as a cumulative percentage series: the percentages retained on each sieve, etc., are summed, progressing from the coarsest to the finest, after the

Table 7.3. *Particle-size distribution, as individual and cumulative percentages, of the sediment in the entrance channel region of the Swanpool Lagoon. Percentages are the mean of five replicate samples. These data are presented graphically in Fig. 8.1*

Mesh aperture of the retaining sieve (mm)	ϕ	Percentage of the total sample	Cumulative percentage
16	−4	2.39	2.39
8	−3	9.81	12.20
4	−2	15.64	27.84
2	−1	25.66	53.50
1	0	20.28	73.78
0.5	1	12.60	86.38
0.25	2	8.20	94.58
0.125	3	3.91	98.49
0.063	4	1.24	99.73
Catch-pot	>4	0.27	100

manner shown in Table 7.3. Secondly, the cumulative percentage series should be plotted against the appropriate ϕ values on arithmetic probability paper (graph paper on which one axis is calibrated in the probability intervals associated with a normal distribution) (see Fig. 7.1). This procedure converts the lognormal curve into something approximating a straight line, making easier the extraction of information. (If arithmetic probability paper is not available, the cumulative percentage series may be plotted against ϕ interval on normal arithmetic graph paper, in which case a sigmoid curve will be obtained.)

Many data useful in the comparison of sediments can be obtained from such graphs, but for most biological purposes measures of average particle size and the degree of scatter about that average will be adequate. These can be obtained by reading off the ϕ values corresponding to certain values of the cumulative percentage; the ϕ value corresponding to 50%, for example, being referred to below as $\phi 50$. Expressions for deriving the median and mean particle sizes and the coefficient of sorting, which quantifies the scattering of the particles about the average size and is equivalent to the standard deviation of the distribution, are:

$$\text{median} = \phi 50$$

$$\text{mean} = \frac{\phi 90 + \phi 80 + \phi 70 + \phi 60 + \phi 50 + \phi 40 + \phi 30 + \phi 20 + \phi 10}{9}$$

$$\text{sorting} = \frac{\phi 90 + \phi 80 + \phi 70 - \phi 30 - \phi 20 - \phi 10}{5.3}.$$

Fig. 7.1. The data in Table 7.3 presented in two forms. *a*. A histogram of the individual percentages of particles falling into the various ϕ categories (note the approximately lognormal distribution). *b*. A plot on arithmetic probability paper of cumulative percentage of particle size against the same ϕ intervals (note the approximately straight line resulting); the line drawn through the points was obtained by regression analysis and has been used to calculate the various characteristics of the sediment shown, for which formulae are given above.

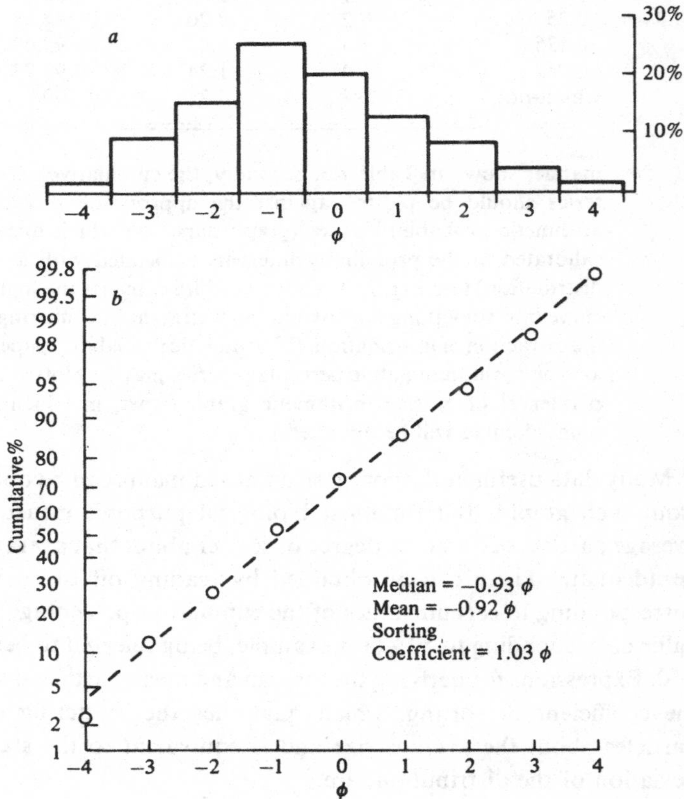

7.2.2 *Characteristics of the water*

Three important characteristics of any lagoonal water mass are its salinity, dissolved oxygen content and transparency: many others, including temperature and pH, can be measured using readily available standard instruments, and the appropriate techniques for quantifying yet others can be obtained from the works listed in Table 7.1.

Water samples for analysis should be obtained from several depths using a piece of equipment such as a Mackereth pump, a very simply

Fig. 7.2. Diagrammatic representation of the Mackereth pump apparatus for obtaining water samples from known depths. After Mackereth, F. J. H., 1963, Freshwater Biological Association Scientific Publication No. 21). *A*, bicycle-tyre pump with the piston washer reversed; *B*, football-inflator valve inserted in rubber tubing the wrong way round; *C*, rubber tubing; *D*, 1.5–2.0-l reservoir which collects water used to rinse out the sample bottle; *E*, screw tap to isolate system whilst the reservoir is emptied; *F*, 250 ml sample bottle (note that the tube through which water enters the bottle extends almost to the bottom); *G*, plastic tubing, graduated in fractions of a metre, and of appropriate length to reach the lagoon bed; *H*, weight to keep tubing '*G*' vertical.

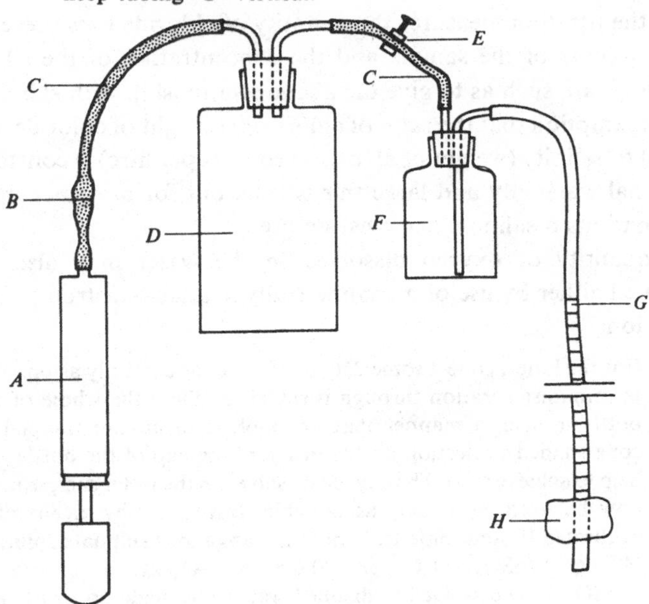

constructed apparatus of which the chief elements are a bicycle-tyre pump and a one-way valve such as that used when inflating a football (see Fig. 7.2). Note (*a*) that the washer on the end of the pump piston must be reversed (concave side facing the handle) so that the pump sucks and does not blow, (b) that the valve must be incorporated the 'wrong way round' so that it prevents air from entering but not from leaving, (*c*) that the reservoir should not be allowed to fill with water (it can be emptied mid-operation whilst the screw-clip is closed), and (*d*) that pumping should be gentle to prevent gas from coming out of solution as a result of the reduced pressure.

Salinity is most easily measured by use of a salinometer of which many types are available commercially. If such an instrument is not available, however, a titration procedure may be used.

> Make up a silver nitrate solution containing 27.25 g per litre and store it in a dark bottle. The salinity of a 10.0 ml water sample can then be found by titrating it with the silver nitrate solution, using 10% potassium chromate as the indicator. At the end point (at which the colour changes from bright yellow to deep orange), the amount of silver nitrate used, in ml, is numerically equivalent to the salinity of the water sample, in ‰.

In fact, the titration measures the quantity of chloride ions present, but the volume of the sample and the concentration of the silver nitrate used are such as to give the above relationship with salinity, on the assumption that the ratio of chlorosity (weight of chloride ion per litre) to salinity (weight of all dissolved salts per litre) is constant in lagoonal water. By and large this is true, but for more accurate determinations a salinometer must be used.

The quantity of oxygen dissolved in the water may also be determined either by use of a commercially-available instrument or by titration.

> For the latter, collect some 250 ml of water as carefully as possible to minimise aeration through turbulence, filling the whole of the bottle in such a manner that no bubbles of air are trapped (a cone-shaped projection on the inside of the cap of the bottle will help to achieve this). The oxygen dissolved in the water sample must now be fixed as quickly as possible. Introduce, by means of a calibrated Pasteur pipette, 2 ml of a manganous sulphate solution (48.0 g $MnSO_4.4H_2O$ *or* 40.0 g $MnSO_4.2H_2O$ *or* 36.4 g $MnSO_4.H_2O$ dissolved in distilled water and made up to 100 ml)

into the bottom of the bottle; then add by a second calibrated Pasteur pipette inserted well below the water level, 2 ml of an alkaline iodide/azide solution. (The alkaline iodide/azide solution is prepared by dissolving 50 g NaOH and 15 g KI in distilled water and making up to 90 ml, then adding 1.0 g NaN_3 dissolved in 4 ml distilled water, and making the whole up to 100 ml.) Replace the cap of the bottle, again ensuring that no air bubbles are trapped, and mix by inverting the bottle several times. Allow the precipitate to settle and shake the bottle again. When the precipitate has again settled, add 2 ml concentrated sulphuric acid, allowing the acid to run down the inside of the bottle, and shake gently until the precipitate has dissolved (if necessary, add more acid). Store the fixed water sample for as short a period as possible before carrying out the titration (see below) and, in any event, keep it out of direct sunlight.

In order to determine its oxygen content, 200 ml of the fixed sample should be titrated with a freshly standardised sodium thiosulphate solution (e.g. dissolve 7.446 g $Na_2S_2O_3.5H_2O$ in freshly-boiled distilled water and make up to 1200 ml with more of the same; add 6 ml chloroform and store in a dark bottle) until a pale straw colour is obtained. Now add 2 ml of soluble starch solution and continue titrating, drop by drop, until the blue colour first disappears. The oxygen dissolved in the water sample, in mg/l, is then given by

$$\frac{\text{ml thiosulphate used} \times 200 \times \text{standardisation factor}^{*}}{\text{volume of water sample, in ml, titrated}}.$$

All wastes from this procedure should be tipped into a special 'poisons container' and not simply flushed down a sink.

The transparency of the water is a factor both affecting and affected by planktonic and submerged plants. It can be estimated very simply by lowering a weighted and white-painted disc of wood or plastic of diameter some 20 cm – a Secchi disc – into the water on the end of a graduated string or pole. The disc is lowered until it can no longer be seen and then it is slowly raised until once again it is just

* The thiosulphate solution may be standardised as follows: place, in a conical flask, 2 g KI dissolved in 100 ml distilled water, 10 ml conc. H_2SO_4, and 20.00 ml of a potassium dichromate solution prepared by dissolving 0.3065 g of oven-dried $K_2Cr_2O_7$ in distilled water and making the volume up to 250 ml; dilute the solution to 200 ml and titrate with the thiosulphate as described above. The standardisation factor is then (20/ml of thiosulphate used).

visible. The depth is read from the string or pole and the process repeated several times to permit a mean value to be calculated.

7.3 Biological sampling

The methodology appropriate to studies of lagoonal organisms varies with the nature of the fauna and flora, and only a few general techniques applicable to most lagoons will be described here. Specialist techniques for specific communities of organisms can be obtained from several of the references listed in Table 7.1.

7.3.1 *Phytoplanktonic primary production*

A relatively simple method for the determination of the primary productivity of the water column, although one not straightforward in its interpretation, relies on the observation that plants liberate oxygen (during photosynthetic production) in the daytime, whilst consuming it (during respiration) throughout the 24-h period.

A stock of translucent glass bottles, of the type used for collecting samples of water for oxygen determinations, is required, of which one in every four should be rendered completely opaque to light. This can be achieved by covering bottle and cap (whilst still permitting the cap to be removed!) with a double layer of black plastic insulating or PVC tape or with aluminium foil or black PVC sheeting. These are then known as the 'dark bottles', as opposed to the untreated 'light bottles'.

By use of the Mackereth pump apparatus (Fig. 7.2 and pp. 81–82), one dark bottle and three light bottles should be filled with lagoonal water from a known depth, observing the same precautions as outlined on p. 82 in respect of water for oxygen determination. Indeed, one of the light bottles should immediately have its contents fixed for later analysis of oxygen content (the procedure described on pp. 82–83 up to and including the dissolution of the precipitate by conc. H_2SO_4). The dark bottle and the two remaining light bottles should then be returned to the same depth in the lagoon from which the water to fill them was drawn. The easiest way of achieving this is to fix them by the neck, so that they project out in a plane parallel to the water surface, to a stout wire itself projecting out sideways from, and attached to, a taut string running from an anchor on the lagoon bed to a surface or subsurface float. Samples from different depths (in multiples of two light and one dark bottles, together with a fourth bottle retained and subjected to the oxygen-fixation treatment) may be collected and attached at the appropriate depth to the same string, provided that they do not overshadow each other and are not shaded by the surface float.

After a period of 24 h or at the end of the daylight period, the bottles should be retrieved from the lagoon and immediately treated to fix the dissolved oxygen. The latter should be measured as soon as possible thereafter as described on p. 83. If the oxygen content, in mg/l, of the water introduced into any given set of four bottles is 'A', that in the light bottles after incubation in the lagoon is 'B' (mean value), and that in the dark bottle is 'C'; then the level of respiration at that depth, per unit volume of water, and over the time period concerned is given by 'A–C', the level of net photosynthesis is given by 'B–A', and of gross photosynthesis by 'B–C'.

Clearly these data are expressed in units of oxygen rather than in terms of quantities of carbon fixed or dry tissue weight produced, the forms in which it is more usual to express productivity. No universal conversion factors can be applied to such oxygen data because of the many other variables involved, but an approximation to amounts of carbon fixed or respired can be obtained by means of the following expressions:

$$\text{gross photosynthesis} = \frac{833.3\dot{}\,(B-C)}{t}\ \text{mgC/m}^3\text{/h}$$

$$\text{net photosynthesis} = \frac{833.3\dot{}\,(B-A)}{t}\ \text{mgC/m}^3\text{/h}$$

and

$$\text{respiration} = \frac{1000(A-C)}{t}\ \text{mgC/m}^3\text{/h}$$

where t is the time, in hours, during which the samples were incubated in the lagoon.

7.3.2 *Organic content of sediments*

Many organisms subsist on organic matter associated with the sediments and therefore a knowledge of the organic content of the bottom sands and muds is often useful. Ideally, one would wish to assess the amounts of the individual components of this organic matter, say in relation to the various consumers, but the apparatus required so to do is both expensive and complex, and as yet we know little of the specific dietary requirements of the relevant species. Hence a crude estimation of total particulate organic matter, including both living and dead components, will be as valuable as any other measure.

One such method, based on incineration of a sediment sample, was given by Barnes (1974) and a second, relying on chemical oxidation (and therefore, strictly speaking, measuring the 'chemical oxygen demand'), will be presented here. The reader should note that it cannot be used for sediments containing appreciable quantities of reduced ions, e.g. anoxic substrata. A sediment sample of known dry weight (about 4 g is generally suitable for muds and organic sands) should be placed with 50 ml of distilled water in a 250 ml narrow-necked conical flask. Then add 10.00 ml of a 1 N potassium dichromate solution, followed by 20 ml of 36 N sulphuric acid (to which has been added 0.0125 g silver sulphate per ml to suppress chloride interference). Shake by hand for one minute and then place in a boiling water bath for half an hour.

When cool, add a further 150 ml of distilled water, 10 ml of 85% phosphoric acid, and 1 ml of diphenylamine indicator solution (0.50 g diphenylamine dissolved in 20 ml distilled water, to which should be added – with caution – 100 ml of 36 N H_2SO_4). The blue-black coloured solution which results should be titrated with 1 N ferrous sulphate until it turns a clear blue. The ferrous sulphate should then be added in aliquots of 0.5 ml until the clear blue colour flashes to green. Add a further 0.50 ml of the dichromate solution and complete the titration, drop by drop, until again the colour flashes from blue to green. If the titration value is less than 1 ml or greater than 9 ml, the analysis should be repeated using one third or three times the quantity of sediment, respectively.

The particulate organic carbon in the sediment sample (the method gives a 75–90% recovery dependent on the nature of the material) is then given by

$$\frac{3(V_1 - V_2)}{W} \text{ mg/g},$$

where V_1 is the volume of dichromate used (10.5 ml), V_2 is that of the ferrous sulphate used, and W is the weight (in g) of the dried sediment sample. Should the ferrous sulphate solution not be exactly 1 N, a standardisation factor will, of course, have to be included to modify the V_2 value. Very roughly, the total organic content of a sediment is equal to 1.8 multiplied by the quantity of organic carbon.

7.3.3 *The macrofauna*

The larger animals (> 0.5 mm) inhabiting lagoons fall fairly naturally into two categories: those of the water mass and those living on or in the sediments and associated vegetation. Planktonic and nektonic species can be sampled, though only semi-quantitatively at best, by means of nets of various mesh sizes, apertures of 0.06–0.09 mm being used for the larger phytoplankton and 0.3 mm for the zooplankton, whilst for the nektonic crustaceans, insects and smaller fish, the author has found a hand-pulled seine-net of 1 mm mesh aperture effective in shallow water. In deeper regions, the same netting can be mounted on an epi-benthic sledge of the type shown in Fig. 7.3 made to a convenient size, or towed in the form of a mid-water trawl.

Fig. 7.3. An epi-benthic sledge used for collecting nekton swimming just above the sediment surface. Wooden uprights (*u*) join together two pairs of moulded fibre-glass skids (*sk*). The two structures so formed are held together by steel strips (*str*), the forward pair of which supports the net (*n*), and are bolted through slots in the uprights, so that the height of the strips above the flat surface of the skids is adjustable. Weights (*w*) are attached to the steel bridle (*b*) behind the pivots (*p*) ensuring that the bridle is always held clear of the sediment. With this arrangement, the sledge tends to tow in the correct attitude before it reaches the bottom. From Barnes *et al*., 1971.

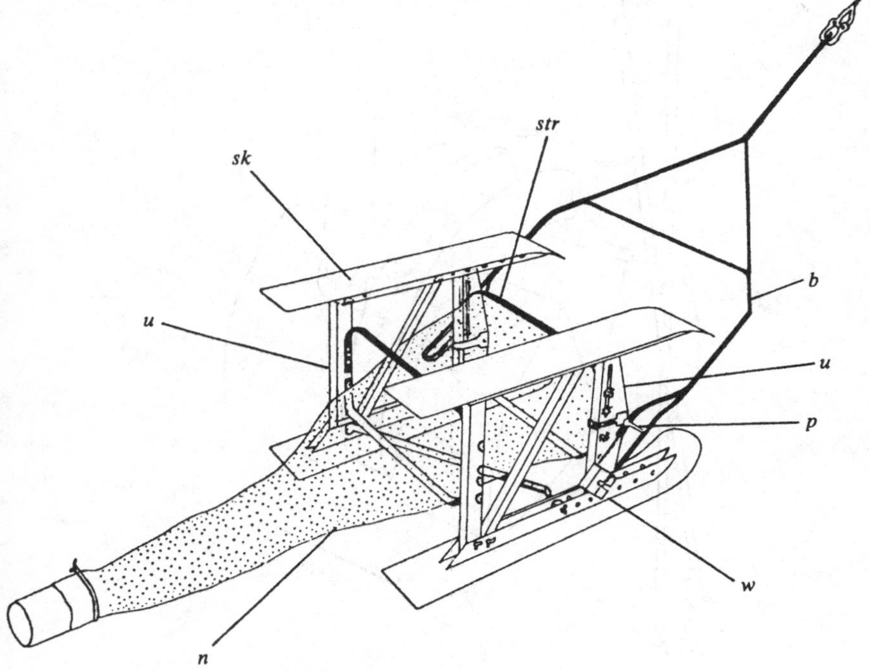

It is easier to collect quantitative information on the benthic fauna, although often more difficult to sort the animals from their surroundings. First it is necessary to take a series of samples of the sediment together with its contained organisms, then to extract all the animals, and finally to sort, examine and quantify the catch. The apparatus required to effect the first two of these steps will vary with the nature of the bottom, with, as before, the distinction being between coarse and fine grained material. The main problem associated with coarse sediments such as gravel and shingle is that many of the inorganic particles are of the same size range or larger than the organisms which it is desired to extract. It is also very difficult to sample sediments supporting dense growths of filamentous algae or submerged macrophytes, except by isolating a sample area in a frame and laboriously picking or pumping out the contents. Special methods will have to be developed for any such vegetated site.

Fig. 7.4. A hand-operated grab for obtaining sediment samples (see text).

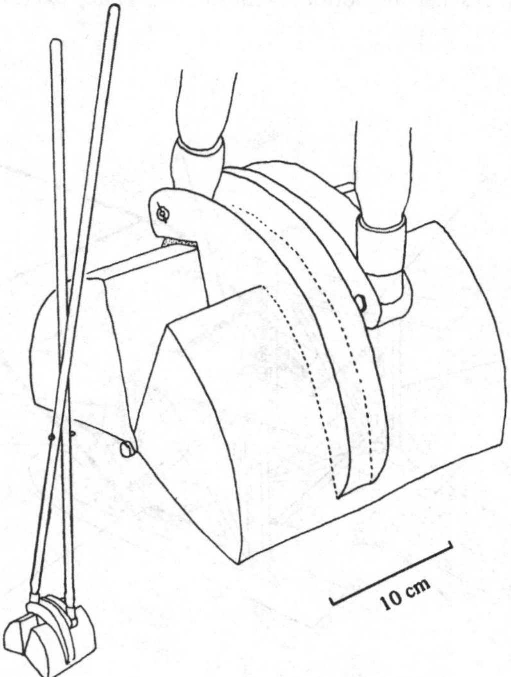

A hand-operated grab of the van Veen type (Fig. 7.4), sampling an area of 0.05 m^2 (e.g. with a square mouth of side length 22.3 cm), will obtain samples of coarse sediments such as those detailed in Fig. 7.1 and Table 7.3. In essence, the grab consists of a fibre-glass hemicylinder, divided into two quarter cylinders which are hinged together along the flat upper surface and which can be opened and closed by means of two stout poles operating on the scissors or tongs principle. The length of the poles can be modified to that appropriate to the depth of water in which one is to operate. The material collected by the grab (that illustrated in Fig. 7.4 bites to a maximum depth of 12 cm) can be discharged on to, and washed through, a coarse sieve (e.g. 5.6–8.0 mm mesh aperture) mounted on top of a large plastic bucket. A sieve and bucket of diameter 30 cm are appropriate to a grab sampling 0.05 m^2. The contents of the bucket are then washed carefully through a sieve of 0.5 mm mesh and the whole contents of that sieve (i.e., sediment plus organisms) preserved and retained for later sorting.

Many gadgets have been designed to extract organisms from sediment but, in the author's experience, none is satisfactory for shingle or gravel. The most effective way of extracting the animals is probably the tedious procedure of tipping all the material retained by the 0.5 mm sieve into a large (e.g. 5 l) polythene beaker, and then agitating it with a controllable and manoeuverable jet of water (e.g. attach a length of rubber tubing to a water tap and partially flatten the end of the tubing between forefinger and thumb when in use). When the beaker is half full of water, decant the suspended material through a 0.5 mm sieve, and repeat the process until two successive cycles yield no further organisms. The sieve then contains all (or most) of the macrofauna associated with the original 0.05 m^2 sample of gravel or shingle.

Samples of finer sediments (sands or muds) can either be obtained with a smaller version of the same grab (e.g. one sampling 0.02 m^2, of side length 14 cm, will suffice for many sands) or with a large core-tube (one of diameter 11.25 cm, sampling 0.01 m^2, will be adequate for most muds). These samples can be tipped straight into the 0.5 mm sieve and 'puddled' in the lagoonal water until the sediment has passed through, leaving the animals (and organic debris) behind. Regardless of sediment type, more than one sample will need to be taken from each sampling site on each visit, the number of samples required depending on the patchiness of the fauna (the sources listed in Table 7.1 will provide guidance).

Whatever the nature of the sediment, one now has the extracted organisms in a 0.5 mm sieve. For counting purposes, etc., it is helpful to wash the contents of the sieve into a rectangular white dish (e.g. of the type used for washing photographic prints) some 30 × 25 cm in size. This can be divided by a grid of black lines into 16–20 equal rectangles to permit samples yielding large numbers of organisms to be subsampled. The various species can then be picked out by hand and counted, weighed, identified, measured, aged, sexed, etc. as appropriate to the study.

By repeating the procedures outlined above at regular intervals, it may be possible to follow changes in the species composition of the fauna, changes in the density and population structure of individual species, changes in species diversity and in many other biological attributes of the lagoon. To obtain a measure of species diversity, for example, one must be able to recognise the presence of different species but one need not necessarily be able to identify them. One diversity index, H, is given by

$$-\Sigma p_i \log_2 p_i,$$

where p_i is the decimal proportion of the total individuals belonging to the ith species. Because $\log_2$ tables are not readily available, it is more convenient to use logs to the base 10, and to rearrange the expression somewhat, so that it becomes

$$H = 3.3219\left(\log_{10} N - \frac{1}{N}\Sigma n_i \log_{10} n_i\right),$$

where N is the total number of individuals of all species, and n_i is the number of individuals of the ith species.

This index can be resolved into two components: that resulting from the number of different species present (s) and that from the equitability with which the total number of individuals present are apportioned between those species (E); i.e. increasing diversity can result either from an increase in s or from a more equitable distribution of individuals amongst the s species. Different environmental factors will affect s and E differently and hence it is often desirable to separate the two components and to ascertain which has the greatest effect in determining the value of H. Clearly, calculation of s is simple, whilst E can be obtained from the expression

$$E = \frac{[\text{antilog}_{10}(0.4343H)] - 1}{s - 1}.$$

Of necessity, this chapter has at times assumed the format of a cookery book, but much valuable and interesting information can also be gained without the aid of any apparatus or complex technique. I would therefore like to conclude by stressing that we know very little of the lives of most lagoonal species. Patient, thorough and informed observation of any species in its natural habitat is likely to increase greatly our understanding of its biology: although 'natural history' is somewhat unfashionable in a number of scientific circles, biology today is perhaps more in need of the observations of the naturalist than it has ever been.

8

Other coastal lagoon-like systems

In the first sentence of this book I pointed out that the term lagoon is used to identify two rather distinct natural features, the coastal lagoon described in the foregoing pages and the lagoon enclosed by a coral reef. The distinction between the two habitats is dramatically emphasised by the fact that their global distributions are almost entirely complementary, and very few regions indeed possess examples of both. The reason for this is simple: the existence of a coastal lagoon is usually dependent on the presence of a reservoir of mobile sand to form the enclosing barrier; mobile sand, however, inhibits the growth of corals. Moving sediment can smother or even sand-blast small, sessile, suspension-feeding animals or those individually capturing small swimming particles, and the larvae of most corals require a firm, rock-like substratum on which to settle. Therefore reefs are characteristic of clear, open, sediment-free waters.

A further difference between the two types of lagoon relates to the extent of separation from the surrounding water mass. Typically, coastal lagoons communicate with the adjacent sea only through a comparatively narrow entrance channel; the barrier separating coral lagoons from the sea, however, is low and is usually overtopped by water during all but the low water of spring tides (see Fig. 8.1). Corals cannot withstand prolonged exposure to the air and they die of desiccation when reef platforms attain heights of some 30 cm above the low water spring tide level. Calcareous algae can build up a slightly higher reef crest, although this still does not exclude the inflow of sea water. Indeed, were water prevented from flowing over the reef crest, the corals and the other reef organisms would no longer be able to obtain sufficient food or nutrients. In addition, water may flow into and out of coral lagoons through the numerous surge channels or passages which dissect the length or circumference of a reef. Thus although coastal lagoons have some claim to be considered separate systems in their own right (the migratory nektonic component notwithstanding), coral lagoons are an integral part of the whole reef ecosystem and cannot really be studied in isolation. A final and more minor difference results from the lack of surface bodies of freshwater on most reefs and atolls; the freshwater input characteristic of many coastal lagoons is therefore absent, except through rainfall, and coral lagoons contain normal sea water (although, in some cases, with a tendency to hyperhalinity). Their fauna and flora are therefore entirely marine in character.

Mention has already been made several times of the difficulty of forcing the continuum of habitat types seen in the natural world into artificial pigeon-holes, and there remain for consideration several habitats which, technically, are not lagoons but which possess a fauna, flora and general biology well within the range shown by

Fig. 8.1. *a*. Sketch-map of a classic Pacific atoll, Kapingamarangi, comprising a reef capped by small islands and encircling a large lagoon. *b*. A diagrammatic section (not to scale) through such an atoll.

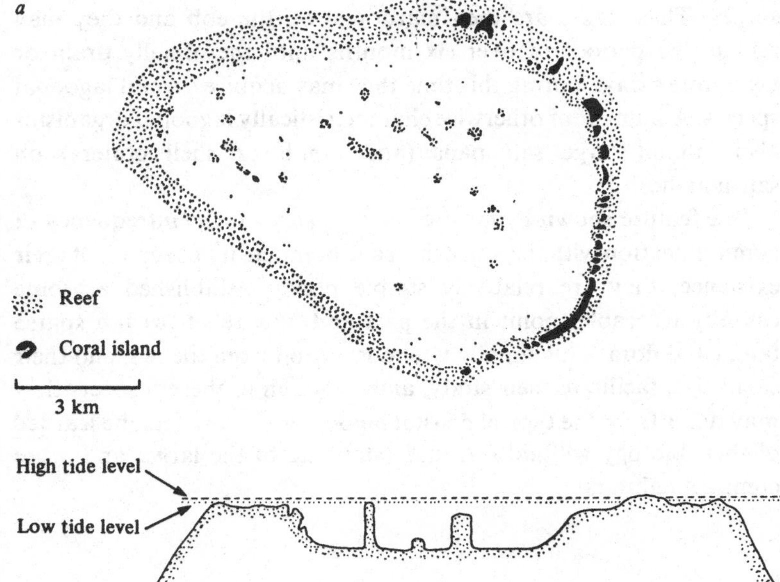

adjacent coastal lagoons. Several of these systems are man-made or man-influenced. Borrow pits and borrow-drains or ditches from which the earth to build sea-walls has been dug form one example, and many drainage ditches, dammed creeks, 'scrapes' constructed for waders and wildfowl, depressions in recently reclaimed land, and relict areas of salt-marsh or mud-flat isolated by coastal defence, embankment or land-reclamation schemes form others. All tend to be fairly small, shallow and short-lived systems, but their abundance around many coasts makes them an ideal subject for investigation by students. On a larger scale, one of the more coastal of the drowned peat-cuttings of Norfolk, Hickling Broad, falls into the same category, and one could arguably include the small ponds dug to maintain and cleanse shellfish, or to function as saltings, or even to serve as bathing pools, and the larger, but otherwise similar, cooling-water ponds of coastal power stations.

Some natural features also show an affinity with coastal lagoons. These are usually associated with large sand-dune systems or, on a smaller scale, with salt-marshes. The high-level hollows between parallel dune ridges – 'lows' or 'slacks' – may be filled with sea water by exceptionally high tides, e.g. during equinoctial tides or storm surges. These lakes are then marooned on the ebb and they may remain for periods of over six months until they finally drain or evaporate away. During this time they may acquire several lagoonal species. A number of otherwise characteristically lagoonal organisms also inhabit large salt pans (and bomb or shell craters) on salt-marshes.

One feature shown by most of these systems is the infrequency of communication with the adjacent sea. For much, if not for all, of their existence, they are relatively simple ponds, established at some (usually traceable) point in the past and now relict, with a sparse fauna and flora. This simplicity and isolation from the sea, and their small size, facilitate their study, and although in these respects they may differ from the typical coastal lagoon, much that can be learned of their biology will aid our understanding of the larger and more complex habitats.

FURTHER READING

BARNES, R. S. K. (1974). *Estuarine Biology*. London: Arnold.

BARNES, R. S. K. & MANN, K. H. (Ed.) (1980). *Fundamentals of Aquatic Ecosystems*. Oxford: Blackwell.

BASSON, P. W., BURCHARD, J. E. Jr, HARDY, J. T. & PRICE, A. R. G. (1977). *Biotopes of the Western Arabian Gulf*. Saudi Arabia: ARAMCO.

BIRD, E. C. F. (1970). *Coasts*. Cambridge, Mass: M.I.T. Press.

CASTAÑARES, A. A. & PHLEGER, F. B. (Ed.) (1969). *Lagunas Costeras, un Simposio*. Mexico: Univ. Nac. Auton. Mexico.

COATES, D. R. (Ed.) (1973). *Coastal Geomorphology*. New York: State University.

COLOMBO, G. (1977). Lagoons. In: BARNES, R. S. K. (Ed.), *The Coastline*. London & New York: Wiley.

EMERY, K. O. (1969). *A Coastal Pond studied by Oceanographic Methods*. New York: Elsevier.

JEFFREY, D. W. (Ed.) (1977). *North Bull Island, Dublin Bay – A Modern Coastal Natural History*. Dublin: Royal Dublin Society.

KNIGHTS, B. & PHILLIPS, A. J. (Ed.) (1979). *Estuarine and Coastal Land Reclamation and Water Storage*. Farnborough: Saxon House.

MCROY, C. P. & HELFFERICH, C. (Ed.) (1977). *Seagrass Ecosystems*. New York: Dekker.

MEE, L. D. (1978). Coastal lagoons. In *Chemical Oceanography*, second edn, volume 7, pp. 441–90. London: Academic Press.

MUUS, B. J. (1967). The fauna of Danish estuaries and lagoons. Distribution and ecology of dominating species in the shallow reaches of the mesohaline zone. *Meddr. Danm. fisk.-og Havunders.* (N.S.), **5**, 1–316.

ODUM, H. T., COPELAND, B. J. & MCMAHAN, E. A. (Ed.) (1974). *Coastal Ecological Systems of the United States*. Washington DC: Conservation Foundation.

PILLAY, T. V. R. (1967). Estuarine fisheries of West Africa (and) Estuarine fisheries of the Indian Ocean coastal zone. In *Estuaries*, ed. G. H. LAUFF, pp. 639–46, 647–57. Washington DC: Amer. Ass. Advmt Sci.

RASMUSSEN, E. (1973). Systematics and ecology of the Isefjord marine fauna. *Ophelia*, **11**, 1–507.

REMANE, A. & SCHLIEPER, C. (1971). *The Biology of Brackish Water*. (*Die Binnengewässer*, 2nd (English) edition, **25**, 1–372.)

RODRIGUEZ, G. (1973). *El Sistema de Maracaibo: Biologia y Ambiente*. Caracas: I.V.I.C.

Sacchi, C. F. (1979). The coastal lagoons of Italy. In *Ecological Processes in Coastal Environments*, ed. R. L. Jefferies & A. J. Davy, pp. 593–601. Oxford: Blackwell.

UNESCO (1979). *Coastal Ecosystems of the Southern Mediterranean: Lagoons, Deltas and Salt Marshes.* Paris: UNESCO. (UNESCO Report in Marine Science No. 7.)

REFERENCES

BARNES, R. S. K. (1974). *Estuarine Biology*. London: Arnold.

BARNES, R. S. K. & HEATH, S. E. (1980). The shingle-foreshore/lagoon system of Shingle Street, Suffolk: a preliminary survey. *Suff. nat. Hist.*, (In press.)

BARNES, R. S. K. & JONES, J. M. (1975). Observations on the fauna and flora of reclaimed land at Calshot, Hampshire. *Proc. Hants Fld Club Archaeol. Soc.*, **29**, 81–91.

BARNES, R. S. K. & MANN, K. H. (Ed.) (1980). *Fundamentals of Aquatic Ecosystems*. Oxford: Blackwell.

BARNES, R. S. K., DOREY, A. E. & LITTLE, C. (1971). An ecological study of a pool subject to varying salinity (Swanpool, Falmouth). An introductory account of the topography, fauna and flora. *J. anim. Ecol.*, **40**, 709–34.

BARNES, R. S. K., WILLIAMS, A., LITTLE, C. & DOREY, A. E. (1979). An ecological study of the Swanpool, Falmouth. IV. Population fluctuations of some dominant macrofauna. In *Ecological Processes in Coastal Environments*, ed. R. L. Jefferies & A. J. Davy, pp. 177–97. Oxford: Blackwell.

BARSDATE, R. J., NEBERT, M. & MCROY, C. P. (1974). Lagoon contributions to sediments and water of the Bering Sea. In *Oceanography of the Bering Sea*, ed. D. W. Hood & E. J. Kelley, pp. 553–76. Fairbanks: University of Alaska.

BAYLY, I. A. E. (1967). The general biological classification of aquatic environments with special reference to those of Australia. In *Australian Inland Waters and their Fauna*, ed. A. H. Weatherley, pp. 78–104. Canberra: ANU Press.

BAYLY, I. A. E. & WILLIAMS, W. D. (1966). Chemical and biological studies of some saline lakes of south-east Australia. *Aust. J. mar. Freshwat. Res.*, **17**, 177–228.

BELLAN, G. (1972). Effects of an artificial stream on marine communities. *Mar. Poll. Bull.*, **3**, 74–7.

BEUKEMA, J. J. (1976). Biomass and species richness of the macro-benthic animals living on the tidal flats of the Dutch Wadden Sea. *Neth. J. Sea Res.*, **10**, 236–61.

BIRD, E. C. F. (1961). The coastal barriers of East Gippsland, Australia. *Geogr. J.*, **127**, 460–8.

BIRD, E. C. F. (1966). The evolution of sandy barrier formations on the East Gippsland coast. *Proc. Roy. Soc. Vict.*, **79**, 75–88.

BIRD, E. C. F. (1970). *Coasts*. Cambridge, Mass.: M.I.T. Press.

BOLTT, R. E. & ALLANSON, B. R. (1975). The benthos of some southern African lakes. III. The benthic fauna of Lake Nhlange, Kwazulu, South Africa. *Trans. Roy. Soc. S. Afr.*, **41**, 241–62.

BOYDEN, C. R. (1972). Relationship of size to age in the cockles *Cerastoderma edule* and *C. glaucum* from the River Crouch Estuary, Essex. *J. Conch.*, **27**, 475–89.

BOYDEN, C. R. (1973). Observations on the shell morphology of two species of cockle *Cerastoderma edule* and *C. glaucum*. *Zool. J. Linn. Soc. Lond.*, **52**, 269–92.

BOYDEN, C. R. & RUSSELL, P. J. C. (1972). The distribution and habitat range of the brackish water cockle (*Cardium* (*Cerastoderma*) *glaucum*) in the British Isles. *J. anim. Ecol.*, **41**, 719–34.

BROCK, V. (1979). Habitat selection of two congeneric bivalves, *Cardium edule* and *C. glaucum* in sympatric and allopatric populations. *Mar. Biol.*, **54**, 149–56.

BROEKHUYSEN, G. J. & TAYLOR, H. (1959). The ecology of South African estuaries. VIII. Kosi Bay Estuary system. *Ann. S. Afr. Mus.*, **44**, 279–96.

CASTAÑARES, A. A. & PHLEGER, F. B. (Ed.) (1969). *Lagunas Costeras, un Simposio*. Mexico: Univ. Nac. Auton. Mexico.

CHAPMAN, C. R. (1971). The Texas Water Plan and its effect on estuaries. In *A Symposium on the Biological Significance of Estuaries*, ed. P. A. Douglas & R. H. Stroud, pp. 40–57. Washington DC: Sport Fishing Institute.

COATES, D. R. (Ed.) (1973). *Coastal Geomorphology*. New York: State University.

COLOMBO, G. (1972). Primi risultati delle ricerche sulle residue Valli di Comacchio e piani delle ricerche future. *Boll. Zool., Italy*, **39**, 471–8.

COLOMBO, G. (1977). Lagoons. In *The Coastline*, ed. R. S. K. Barnes, pp. 63–81. London & New York: Wiley.

COMIN, F. A. & FERRER, X. (1978). Desarrollo masivo del fitoflagelado *Prymnesium parvum* Carter (Haplophyceae) en una laguna costera del delta del Ebro. *Oecologia aquatica*, **3**, 207–10.

CRAWFORD, R. M., DOREY, A. E., LITTLE, C. & BARNES, R. S. K. (1979). Ecology of the Swanpool, Falmouth, V. Phytoplankton and nutrients. *Est. coast. mar. Sci.*, **9**, 135–60.

CROMWELL, J. E. (1971). Barrier coast distribution: a world-wide survey. *Abstr. Vol. 2nd Nat. Coast. Shallow Water Res. Conf.*, p. 50.

DOREY, A. E., LITTLE, C. & BARNES, R. S. K. (1973). An ecological study of the Swanpool, Falmouth, II. Hydrography and its relation to animal distribution. *Est. coast. mar. Sci.*, **1**, 153–76.

EDWARDS, R. R. C. (1977). Field experiments on growth and mortality of *Penaeus vannamei* in a Mexican coastal lagoon complex. *Est. coast. mar. Sci.*, **5**, 107–21.

EDWARDS, R. R. C. (1978). Ecology of a coastal lagoon complex in Mexico. *Est. coast. mar. Sci.*, **6**, 75–92.

EMERY, K. O. (1969). *A Coastal Pond Studied by Oceanographic Methods*. New York: Elsevier.

FENCHEL, T. (1975*a*). Factors determining the distribution patterns of mud snails (Hydrobiidae). *Oecologia, Berl.*, **20**, 1–17.

FENCHEL, T. (1975*b*). Character displacement and coexistence in mud snails (Hydrobiidae). *Oecologia, Berl.*, **20**, 19–32.

FENCHEL, T. (1977). Aspects of the decomposition of seagrasses. In *Seagrass Ecosystems*, ed. C. P. McRoy & C. Helfferich, pp. 123–45. New York: Dekker.

FENCHEL, T. & KOFOED, L. H. (1976). Evidence for exploitative interspecific competition in mud snails (Hydrobiidae). *Oikos*, **27**, 367–76.

FERRER, X. & COMIN, F. A. (1979). Distribucio i ecologia dels macrofits submergits del delta de l'Ebre. *Butll. Inst. catal. Hist. nat.*, **44**, 111–17.

Fiala, M. (1973). Études physico-chimiques des eaux et sédiments de l'Etang Bages-Sigean (Aude). *Vie Milieu*, **23**, 21–50.

Frier, J. O. (1979). Character displacement in *Sphaeroma* spp. (Isopoda; Crustacea). I. Field evidence. II. Competition for space. *Mar. Ecol. Prog. Ser.*, **1**, 159–68.

Gierloff-Emden, H. G. (1961). Nehrungen und Lagunen. *Petermanns Geogr. Mitt.*, **105**, 81–92, 161–76.

Gilmartin, M. & Revelante, N. (1978). The phytoplankton characteristics of the barrier island lagoons of the Gulf of California. *Est. coast. mar. Sci.*, **7**, 29–47.

Hedgpeth, J. W. (1953). An introduction to the zoogeography of the northwestern Gulf of Mexico with reference to the invertebrate fauna. *Publ. Inst. mar. Sci., Texas*, **3**, 107–224.

Hedgpeth, J. W. (1957). Estuaries and lagoons. II. Biological aspects. *Geol. Soc. Amer. Mem.* No. 67, 693–729.

Hedgpeth, J. W. (1967). Ecological aspects of the Laguna Madre, a hypersaline estuary. In *Estuaries*, ed. G. H. Lauff, pp. 408–19. Washington, DC: Amer. Ass. Advmt Sci.

Heip, C. & Herman, R. (1979). Production of *Nereis diversicolor* O. F. Muller (Polychaeta) in a shallow brackish water pond. *Est. coast. mar. Sci.*, **8**, 297–305.

Hellier, T. R. Jr (1962). Fish production and biomass studies in relation to photosynthesis in the Laguna Madre of Texas. *Publ. Inst. mar. Sci., Texas*, **8**, 1–22.

Hill, B. J. (1975). The origin of southern African coastal lakes. *Trans. Roy. Soc. S. Afr.*, **41**, 225–40.

Hill, M. B. & Webb, J. E. (1958). The ecology of Lagos Lagoon II. The topography and physical features of Lagos Harbour and Lagos Lagoon. *Phil. Trans. Roy. Soc.*, **241**, 319–33.

Hodgkin, E. P. (1976). The history of two coastal lagoons at Augusta, Western Australia. *J. Roy. Soc. W. Aust.*, **59** (2), 39–45.

Holm, R. F. (1978). The community structure of a tropical marine lagoon. *Est. coast. mar. Sci.*, **7**, 329–45.

Hopkinson, C. S., Gosselink, J. G. & Parrondo, R. T. (1978). Aboveground production of seven marsh plant species in coastal Louisiana. *Ecology*, **59**, 760–9.

Howard-Williams, C. (1978). The growth and production of aquatic macrophytes in a south temperate saline lake. *Verh. Internat. verein. Limnol.*, **20**, 1153–8.

Howard-Williams, C. & Davies, B. R. (1979). The rates of dry matter and nutrient loss from decomposing *Potamogeton pectinatus* in a brackish south-temperate lake. *Freshwat. Biol.*, **9**, 13–21.

Hunt, O. D. (1971). Holkham Salts Hole, an isolated salt-water pond with relict features. An account based on studies by the late C. F. A. Pantin. *J. mar. biol. Ass. U.K.*, **51**, 717–41.

Hylleberg, J. (1975). The effect of salinity and temperature on egestion in mud snails (Gastropoda: Hydrobiidae). *Oecologia, Berl.*, **21**, 279–89.

Jacques, G., Cabet, G., Fiala, M., Neveux, J. & Panousse, M. (1975). Caractéristiques du milieu pélagique des Étangs de Bages-Sigean et de Salses-Leucate pendant l'été 1974. *Vie Milieu*, **25**, 1–18.

Jones, D. A., Price, A. R. G. & Hughes, R. N. (1978). Ecology of the high saline lagoons Dawhat as Sayh, Arabian Gulf, Saudi Arabia. *Est. coast. mar. Sci.*, **6**, 253–62.

KAPLIN, P. A. (1959). (Some features of the lagoons along the northeastern shores of the USSR.) (In Russian.) *Trudȳ okeanogr. Kom.*, **4**, 54–65.

KIKUCHI, T. & PÉRÈS, J. M. (1977). Consumer ecology of seagrass beds. In *Seagrass Ecosystems*, ed. C. P. McRoy & C. Helfferich, pp. 147–93. New York: Dekker.

KIKUCHI, W. K. (1976). Prehistoric Hawaiian fishponds. *Science, N.Y.*, **193**, 295–9.

KING, C. A. M. (1972). *Beaches and Coasts*. 2nd Ed. London: Arnold.

LANKFORD, R. R. (1977). Coastal lagoons of Mexico. Their origin and classification. In *Estuarine Processes*, Volume 2, ed. M. Wiley, pp. 182–215. New York: Academic Press.

LANKFORD, R. R. (1978). Man's use of coastal lagoon resources. In *Advances in Oceanography*, ed. H. Charnock & G. Deacon, pp. 245–53. New York: Plenum.

LASSEN, H. H. (1979). Reproductive effort in Danish mudsnails (Hydrobiidae). *Oecologia, Berl.*, **40**, 365–9.

LASSEN, H. H. & CLARK, M. E. (1979). Comparative fecundity in three Danish mudsnails (Hydrobiidae). *Ophelia*, **18**, 171–8.

LEONT'EV, O. K. & LEONT'EV, V. K. (1957). (The origin and developmental features of lagoon coasts.) (In Russian.) *Trudȳ okeanogr. Kom.*, **2**, 86–103.

LEVINTON, J. S. (1972). Stability and trophic structure in deposit-feeding and suspension-feeding communities. *Amer. Nat.*, **106**, 472–86.

LITTLE, C., BARNES, R. S. K. & DOREY, A. E. (1973). An ecological study of the Swanpool, Falmouth. 3. Origin and history. *Cornish Stud.*, **1**, 33–48.

MCROY, C. P. & MCMILLAN, C. (1977). Production ecology and physiology of seagrasses. In *Seagrass Ecosystems*, ed. C. P. McRoy & C. Helfferich, pp. 53–87. New York: Dekker.

MANN, K. H. (1976). Decomposition of marine macrophytes. In *The Role of Terrestrial and Aquatic Organisms in Decomposition Processes*, ed. J. M. Anderson & A. Macfadyen, pp. 247–67. Oxford: Blackwell.

MARCHESONI, V. (1954). Il trofismo della Laguna Veneta e la vivicazioni marina. III. Ricerche sulle variazioni quantitative del fitoplankton. *Arch. Oceanogr. Limnol.*, **9**, 147–281.

MEE, L. D. (1978). Coastal lagoons. In *Chemical Oceanography*, 2nd edn, volume 7, pp. 441–90. London: Academic Press.

MUUS, B. J. (1967). The fauna of Danish estuaries and lagoons. Distribution and ecology of dominating species in the shallow reaches of the mesohaline zone. *Meddr. Danm. fisk.-og Havunders.* (N.S.), **5**, 1–316.

ORME, A. R. (1973). Barrier and lagoon systems along the Zululand coast, South Africa. In *Coastal Geomorphology*, ed. D. R. Coates, pp. 181–217. New York: State University.

PARKER, M. M. & WEST, A. B. (1979). The natural history of *Neomysis integer* (Leach) in Lough Furnace, Co. Mayo, a brackish lough in the west of Ireland. *Est. coast. mar. Sci.*, **8**, 157–67.

PETERSON, C. H. (1975). Stability of species and of community for the benthos of two lagoons. *Ecology*, **56**, 958–65.

PILLAY, T. V. R. (1967). Estuarine fisheries of West Africa (and) Estuarine fisheries of the Indian Ocean coastal zone. In *Estuaries*, ed. G. H. Lauff, pp. 639–46, 647–57. Washington, DC: Amer. Ass. Advmt Sci.

Por, F. D. (1972). Hydrobiological notes on the high salinity waters of the Sinai Peninsula. *Mar. Biol.*, **14**, 111–19.

Qasim, S. J. (1970). Some problems related to the food chain in a tropical estuary. In *Marine Food Chains*, ed. J. H. Steele, pp. 45–51. Edinburgh: Oliver & Boyd.

Rasmussen, E. (1973). Systematics and ecology of the Isefjord marine fauna. *Ophelia*, **11**, 1–507.

Remane, A. & Schlieper, C. (1971). *The Biology of Brackish Water.* (*Die Binnengewässer*, 2nd (English) edn, **25**, 1–37.)

Sacchi, C. F. (1979). The coastal lagoons of Italy. In *Ecological Processes in Coastal Environments*, ed. R. L. Jefferies & A. J. Davy, pp. 593–601. Oxford: Blackwell.

Salemaa, H. (1979). Ecology of *Idotea* spp. (Isopoda) in the northern Baltic. *Ophelia*, **18**, 133–50.

Schachter, D. (1969). Ecologie des eaux saumâtres. *Verh. Internat. verein. Limnol.*, **17**, 1052–68.

Schwartz, M. L. (Ed.) (1973). *Barrier Islands*. Stroudsburg: Dowden, Hutchinson & Ross.

UNESCO (1979). *Coastal Ecosystems of the Southern Mediterranean: Lagoons, Deltas and Salt Marshes.* Paris: UNESCO (UNESCO Report in Marine Science No. 7.)

Van Valen, L. (1971). Group selection and the evolution of dispersal. *Evolution*, **25**, 591–8.

Vatova, A. (1960). Primary production in the High Venice Lagoon. *J. Conseil*, **26**, 148–55.

Walsh, G. E. (1965). Studies on dissolved carbohydrate in Cape Cod waters. II. Diurnal fluctuation in Oyster Pond. *Limnol. Oceanogr.*, **10**, 577–82.

Warburton, K. (1978). Community structure, abundance and diversity of fish in a Mexican coastal lagoon system. *Est. coast. mar. Sci.*, **7**, 497–519.

Webb, J. E. (1958). The ecology of Lagos Lagoon I. The lagoons of the Guinea coast. *Phil. Trans. Roy. Soc.*, **241**, 307–18.

White, D. A., Weiss, T. E., Trapani, J. M. & Thien, L. B. (1978). Productivity and decomposition of the dominant salt marsh plants in Louisiana. *Ecology*, **59**, 751–9.

Whittaker, R. H. (1975). *Communities and Ecosystems*, 2nd edn. New York: Macmillan.

Williams, R. B. (1973). The significance of saline lagoons as refuges for rare species. *Trans. Norf. Norw. nat. Soc.*, **22**, 387–92.

Wøldike, K. (1973). Phytoplankton in the oligohaline lake, Selsø. Primary production and standing crop. *Ophelia*, **12**, 27–44.

Wood, E. J. F., Odum, W. E. & Zieman, J. C. (1969). Influence of sea grasses on the productivity of coastal lagoons. In *Lagunas Costeras, un Simposio*, ed. A. A. Castañares & F. B. Phleger, pp. 495–502. Mexico: Univ. Nac. Anton. Mexico.

Young, D. K. & Young, M. W. (1977). Community structure of the macrobenthos associated with seagrass of the Indian River Estuary, Florida. In *Ecology of Marine Benthos*, ed. B. C. Coull, pp. 359–82. Columbia: Univ. S. Carolina Press.

Zenkevitch, L. (1963). *Biology of the Seas of the USSR.* London: Allen & Unwin.

ZENKOVICH, V. P. (1967). *Processes of Coastal Development*. Edinburgh: Oliver & Boyd.

ZIMMERMAN, R., GIBSON, R. & HARRINGTON, J. (1979). Herbivory and detritivory among gammaridean amphipods from a Florida seagrass community. *Mar. Biol.*, **54**, 41–7.

INDEX

Lagoons and geographical regions

INDEX

Organisms